FACTORS INFLUENCING CLOUD COMPUTING ADOPTION BETWEEN GOVERNMENT, ACADEMIC, AND FOR-PROFIT CORPORATE ORGANIZATIONS

A CORRELATIONAL STUDY

Dr. Sean C. Lawrence

A Dissertation Presented in Partial Fulfillment
of the Requirements for the Degree
Doctor of Management

First Edition

PAGE PUBLISHING
Conneaut Lake, PA

First originally published by Page Publishing 2024

ISBN 979-8-89315-054-4 (pbk)
ISBN 979-8-89315-078-0 (digital)

Printed in the United States of America

University of Phoenix Dissertation Committee for Sean Lawrence certifies approval of the following dissertation:

Factors Influencing Cloud Computing Adoption between Government, Academic, and For-Profit Corporate Organizations

A Correlational Study

Committee:

Douglas Neeley, PhD, Chair
Hakeem Lumumba, PhD, Committee Member
Elizabeth Young, EdD, Committee Member

____

Douglas Neeley

____

Hakeem Lumumba

____

Elizabeth Young

____

Hinrich Eylers, PhD
Vice Provost, Doctoral Studies
University of Phoenix

Date Approved: ______________

This dissertation is dedicated first and foremost to my late mother, Lorna Richardson, who, for as long as I can remember, constantly instilled in me the drive to always move forward in a progressive and positive manner. You pushed me to heights I did not know I could reach. In your memory, I strove to attain the one thing you told me nobody could ever take from me—my education.

To my family who have been on this journey with me from day one, thank you for standing by me. Through the ups, downs, trials, and tribulations encountered along this journey, your love, support, and continued encouragement gave me the strength and fortitude to press forward when my days were dark and my nights were darker.

This dedication is more especially for my children, Tony, Maliq, and TeAhna. Watching the three of you grow up has been the greatest joy of my life. To my sons, you are the men that I wish I could have been. The drive you young men have in the pursuit of excellence is nothing short of amazing. To my little princess, your unselfish and loving spirit amazes me every time I look at you. Your hugs and kisses always made my day look a little brighter. But when you would whisper to me, "Daddy, you got this, I love you," that was the fuel that got me over the edge.

CONTENTS

LIST OF FIGURES

LIST OF TABLES

ACKNOWLEDGMENTS

First and foremost, I want to give thanks and praise to the Grand Architect of the Universe, our Lord and Savior Jesus Christ, who gave me the mental fortitude and the spiritual resolve to make it through this doctoral journey. I wish to thank my dissertation committee members who provided me constructive feedback along the way to help me successfully get through this process. But most especially to Dr. Lumumba for your responsiveness and being the voice of reason when I was starting to panic. To Dr. Liz for staying up late and working with me for almost three hours on perfecting my dissertation.

The list of family and friends who kept me encouraged and sent prayers up for me is extensive extensive, so to all of you, I say a heartfelt Thank You. To my fellow doctoral classmates—Donte and Leslie—who refused to let me quit and often had to talk me "off the ledge." To all my coworkers who supported me from the day they found out I was pursuing a doctoral degree, telling me they couldn't wait to call me Dr. Lawrence. To my boss, Bob Capps, who was so supportive and gave me a flexible schedule that allowed me the time needed to accomplish my homework assignments while still being able to support the war fighter.

Finally, most of all, I want to thank all the naysayers, the haters, and all the people who told me I couldn't do this and would fail. Your negativity did nothing but strengthen my resolve to finish this program and obtain my doctoral degree in spite of your desires for me to fail. How ya like me now?

ABSTRACT

The success of the Internet over the past decade has been the catalyst for the emergence of new technologies like cloud computing. The evolution of cloud computing has promoted innovation in goods and services while providing competitive market advantages to various organizations across numerous industries. Cloud computing has become one of the foremost technology strategies business, government, and educational organizations consider to extend their infrastructure, enhance their growth opportunities, and gain new capabilities. However, seminal research found decision-makers did not understand the factors that contributed to the decision to adopt cloud computing. To achieve the purpose of this study, a quantitative nonexperimental correlational research method was used to determine the relationship between the factors that influenced IT directors' decision to adopt cloud-computing technology within their respective for-profit corporate entities, academic institutions, and government agencies. The analysis of data resulted in a statistically significant correlation between the security effectiveness, cost savings, and reliability factors and the decision to adopt cloud computing in government organizations, for-profit corporate organizations, and academic institutions. However, between the organizations, the study revealed the decision to adopt cloud computing was significantly higher in for-profit corporate organizations than for government or academic institutions.

1

Introduction

The success of the Internet over the past decade has been the catalyst for the emergence of new disruptive technologies like cloud computing. The evolution of cloud computing has transformed traditional information technology (IT) business models from offering IT as a product to offering IT as a service. Habjan and Pucihar (2017) found advancements in the Internet, coupled with cloud computing, have proven to increase innovation in goods and services while providing competitive market advantages to enterprises across various sectors. Cloud computing has become one of the foremost technology strategies business, government, and educational organizations consider to extend their infrastructure, enhance their growth opportunities, gain new capabilities, and launch new innovations without having to procure additional hardware or software (Akintomide 2013).

The use of cloud computing has become an integral part of business as the benefits in cost, flexibility in deployment, gains in efficiency, and resource pooling provide distinct business advantages to the organization (Ibraham 2014). For many organizations, the introduction of cloud computing has been the driving force behind the creation of numerous jobs across various business units while also becoming the catalyst for revenue and an enabler for new business models (Tripathi and Jigeesh 2013).

Cloud computing has altered the nature of the relationship between an organization and its customers. Leveraging cloud technologies, an organization is better poised to provide a flexible system that seamlessly links the company with their customers via mobile platforms to provide service around the world at any time of day; however, despite these benefits, cloud computing adoption has been slow (Asadi, Nilashi, Abd Razak, and Yadegaridehkordi 2017). Organizations recognize new technologies such as cloud computing have become a strategic resource, and the investment into such technology should bear tangible and quantifiable benefits to justify the investment (Chen, Ta-Tao, and Kazuo 2016). In addition to the benefits and advantages cloud computing has made in traditional for-profit corporate organizations, cloud technologies offer government agencies solutions for automating and streamlining civic processes (Daniel 2018).

As technology continues to advance, government organizations are under constant pressure to innovate in order to meet the rising demands and expectations of their citizen population. Implementing cloud technology yields operational models capable of providing a single gateway to government-related services and data (Kaur 2018). When considering cloud adoption within the government sector, trust is a critical component. Unfortunately, because cloud computing is a relatively new and emerging technology with an underdeveloped marketplace for cloud services, implementation within government agencies has proven to be challenging (Nikolova 2012). When implemented correctly in a government agency, cloud computing can improve transparency between citizens, business owners, and government administrators while improving communication and collaboration across the organization (Nikolova 2012).

According to Nikolova (2012), citizen services are improved through the innovation of data services in the cloud. Government agencies can reduce the footprint of their data centers by only paying for the resources needed, adding a degree of flexibility where resources can be scaled up or down to meet near-term requirements (Nikolova

2012). In comparison to other sectors, government agencies have a pivotal responsibility in facilitating the adoption of cloud computing, according to Kaur (2018), policymakers and government officials established and implemented digital government decisions that specifically addressed cloud computing.

Academia has a vital role in the economic growth of an organization and a country. The traditional classroom has undergone a metamorphosis. As the technological landscape continues to change and colleges and universities continue to face reduced operating budgets, institutions of higher learning are forced to adopt new technologies that enable their e-learning systems to keep pace with technology while reducing costs and staying within federal compliance regulations (Saini and Kauer 2017). Different authors have different points of view on the cloud in education. Rajaraman (2014a) believed educational institutions have to be concerned with the security of their programs and data, the challenges of moving their applications to a different service provider if the quality of service declines or the cloud service provider ceases operations, and the distribution of their data across national borders. However, Muniasamy, Ejalani, and Anandhavalli (2014) deduced that institutions of higher learning could take advantage of cloud computing's security, flexibility, portability, efficiency, and reliability.

Background of the Problem

In 1965, the cofounder of the semiconductor chip company Intel, Dr. Gordon Moore, made the assertion that the number of transistors on a chip would double every eighteen to twenty-four months; today, technology advances daily (Li 2016 and Mollick 2006). Typically, the advancements are on systems; however, there are occasions when technological advancements deliver a new way of solving an old problem. To ensure survivability in twenty-first-century global markets, organizations must take advantage of new technologies that facilitate the development and delivery of higher-quality services.

Corporate entities, academic institutions, and government agencies invest a significant amount of their operating budget into information systems to assist their respective organizations in accomplishing their operational goals by supporting their strategic objectives while improving employee productivity and providing the organization a competitive advantage. Corporate entities, academic institutions, and government agencies have traditionally been burdened with maintaining the rising costs of their IT infrastructure and data centers that support networking equipment, servers, and storage arrays (English 2018).

Cloud services as an emerging paradigm deliver the full spectrum of IT services on demand as a flexible and scalable option with a considerable cost saving over traditional service offerings (Wang, He, and Wang 2012). Research into cloud computing over the years has yielded salient information as such numerous definitions have been associated with cloud computing. According to Ji-Seong Kim and Kwan-Hee (2013), Google pioneered the term *cloud computing* and defined it as an IT service that uses the Internet to deliver services and capability. Cloud platforms can provide a model for access to business applications as well as the storing, processing, and sharing of data via the web (Rzheuskiy, Veretennikova, Kunanets, and Kut 2018). Under the cloud model, the responsibility for providing compute resources as well as the underlying infrastructure are assumed by cloud service providers.

Cloud computing business service models are divided into three primary categories: Software as a Service (SaaS), Platform as a Service (PaaS), and Infrastructure as a Service (IaaS) (Iqbal, Kiah, Anuar, Daghighi, Wahab, and Khan 2016).

As a service model, SaaS institutes processes and applications that provide customers access to services within the cloud environment. Using the SaaS service model, customers are no longer burdened with having to manage the underlying infrastructure as that responsibility is shifted to the cloud service provider (Iqbal et al. 2016).

PaaS is a model that provides an organization the flexibility to rent cloud services, which incorporate configurable application environments consisting of hardware, operating systems, applications, and databases to host web applications and perform software development (Iqbal et al. 2016 and Krancher, Luther, and Jost 2018). Software development teams are the principal consumers of the PaaS offering as seminal research conducted by Krancher et al. (2018) indicated development teams experienced a 30 percent increase in shorter development cycles after adopting the PaaS model. Unlike SaaS, PaaS offers the features of shaping environments through rapid elasticity and abstraction, the reuse of software services, the enablement of self-organization, and a mechanism for obtaining continuous feedback (Krancher et al. 2018).

IaaS provides an organization with the ability to provision networking, processing, storage, and server resources using virtualization technology that allows for resource pooling in a multitenant environment (Kim 2017). The IaaS construct enables organizations to completely outsource their data center as opposed to maintaining the underlying infrastructure typically required to host the company's information; the resulting cost benefits are realized over time as customers only pay for the resources used (Krancher et al. 2018).

Problem Statement

The general focus of this doctoral dissertation study was to discover if organizational type influenced the adoption of cloud computing among organizations in Central Florida. While seminal research into cloud computing has been published, Gill (2011), Hus and Lin (2016), and Low, Chen, and Wu (2011) found many decision-makers did not understand the factors that contributed to the adoption of cloud computing. Additionally, while there was increased interest among senior decision-makers in adopting cloud computing, concerns over the technology's reliability, stability, and security prevented prompt decision-making (Marston, Li, Bandyopadhyay, Zhang, and Ghalsasi 2011).

Specific Problem

The specific problem is when institutions blindly adopt cloud computing, the result is a lack of understanding of the different factors that influence the adoption decision. Research conducted by Alfifi (2015) evaluated predetermined criteria of CIOs of higher education institutions in relation to their perceptions of the security effectiveness, cost saving, reliability, and accountability of cloud computing, while Hakim (2017) conducted research with CIOs of technology- and nontechnology-oriented companies that focused on the factors that contributed to the resistance of the adoption of cloud computing.

Previous research conducted by Nikolova (2012) revealed the use of cloud computing to be more popular with business organizations and less favored within government agencies. Traditionally, federal government agencies are risk averse to adopting cloud computing technology, citing the newness and relative underdevelopment of the cloud services marketplace (Nikolova 2012). However, Daniel (2018) found utilizing cloud technology, state and local government agencies were able to simplify operations while allowing them to improve service to citizens, respond to developing technology needs, and better manage cybersecurity issues.

Purpose of the Study

The purpose of this correlational study was to discover the relationship between the key factors (security effectiveness, cost savings, and reliability) that influence IT directors' decisions to adopt cloud-computing technology within their respective institutions. According to Ghazizadeh, Lee, and Boyle (2012), previous research revealed the Technology Acceptance Model (TAM) has been used and validated across various technologies and populations; as such, the TAM was used as a guide to identify the internal and external factors impacting the decision on whether or not to adopt cloud computing. The results of this study provided salient information that contributed to the body of knowledge in cloud computing by informing decision-makers of viable approaches they can employ to increase acceptance of cloud computing within their respective organizations.

Population and Sample

The target population for this study was IT directors or equivalents of government agencies, for-profit corporate organizations, and academic institutions of higher education in Central Florida who are the principal decision-makers in IT procurement decisions. To obtain the appropriate sample size for this study power analysis, alpha level = .05, effect size = .30, power = .80, and number of groups = 3 with a one-way ANOVA will be used, yielding a total sample population of 111 participants (37 participants per group). The concept of power analysis refers to the probability of rejecting a false null hypothesis, while the alpha level is the percentage of time the null hypothesis will be incorrectly rejected, and the effect size measures the size of the relationship between the independent and dependent variable (Christen, Johnson, and Turner 2014; Meyvis and Van Osselaer 2018). While it would have been beneficial to collect data from the entire targeted population, Christen, Johnson, and Turner (2014) posited it was impractical and unnecessary to use too large a population. The purposive sampling method was used to identify participants meeting the criteria of being a principal decision-maker in IT procurement decisions within their respective organizations. In purposive sampling, participants are chosen based on predetermined criteria established by the researcher (Leedy and Ormrod 2016).

Significance of the Study

Cloud computing has emerged as the latest disruptive technology that offers the delivery of information technology infrastructure, platform, and software to organizations as a pay-as-you-go service. While cloud computing has provided opportunities for organizations to increase productivity while reducing costs, the technology has caused many organizations significant concern. As with any new technology implementation, in addition to start-up costs such as changes to the organization's infrastructure, organizational change must also be considered. Organizational change will consist of actions that influence what people within the organization do, how they perform their day-to-day tasks, and how their roles and responsibilities are defined as cloud computing is adopted within the organization (Brazier 2014). Some of the most complicated changes an organization can undergo will involve IT implementations (Krogh 2018). In a time when organizations are faced with doing more with less but must remain responsive to customer demands while maintaining a competitive edge in the marketplace, implementing new IT initiatives and processes can offer an organization a myriad of benefits. However, for an organizational change involving IT to be successful, the organization must be prepared to receive and adapt to technology.

Significance for Research

The research study added to the existing body of knowledge concerning the adoption of cloud-computing technology in the context of government agencies, for-profit businesses, and academic institutions in Central Florida. The research built upon previous research conducted on the adoption of cloud computing by Alfifi (2015); Marković, Branović, and Popović (2014); and Nikolova (2012) by utilizing a similar research design employed by each of the respective researchers, however, using a different population. From a research perspective, while cloud computing appears to offer several benefits and advantages to organizations that decide to adopt the technology, Rajendra Prasad, Lakshaman Naik, and Bapuji (2013) warn there are still several challenges researchers and academicians need to be aware of as it pertains to security, reliability, and privacy. The results of this study will address which factor (security, cost, or reliability) was statistically significant between the three groups evaluated as part of this study in the decision to adopt cloud computing.

The research used the TAM as the theoretical framework underpinning this study. According to Alfifi (2015), the TAM has proved to be an effective model in predicting users' attitudes toward new technologies. The theoretical significance of this research will be recognized as this study further validated and tested the TAM against a different sample population to determine if the TAM is an effective model to understand the relationship between the factors that influence IT directors' decisions to adopt cloud-computing technology within their respective institutions. This study provided practical value to IT directors in government agencies who can use the research findings to aid in the development of acceptance criteria for services delivered by third-party service providers. Additionally, the research findings can be used as a reference for the establishment of service-level agreements between government agencies and cloud service providers to ensure the proper protection of government data.

IT professionals in academic institutions will be able to use the research findings to work with administrators to develop and design new curriculums for digital delivery. The findings would enable the discovery of innovative ways to deliver the virtual classroom while improving the student educational experience.

Furthermore, the results of this study would bring awareness of the advantages and disadvantages of adopting cloud-computing technologies within a business organization. Using the results, IT directors or equivalents could prioritize the factors that are important and relevant to their respective business organizations.

There are a number of leadership implications the technology community will encounter as the study findings will highlight the need for leaders within the industry to address issues with governance. The results of the study may be useful for industry leaders to not only make decisions but to recommend changes to existing policies. According to Al-Ruithe and Benkhelifa (2017), establishing a data governance strategy in cloud environments is one of the biggest challenges organizations will face. Civic leaders in both federal and state government may use the study results to close the gaps in current security and data protection policies for cloud-computing service providers.

Nature of the Study

The goal of this research study was to discover the relationship between the factors that influenced IT directors' decisions to adopt cloud-computing technology within their respective institutions. To achieve the intended purpose of this study, a quantitative approach was used. Cooper and Schindler (2014) discussed the quantitative methodology as being appropriate when the researcher seeks to measure behavior or attitudes. Additionally, quantitative methods focus on describing, explaining, or predicting a phenomenon by testing or building upon an existing theory as this study uses the TAM as its theoretical foundation (Cooper and Schindler 2014).

In contrast, the qualitative research methodology seeks to provide an in-depth understanding of a particular phenomenon and is traditionally used to inform the researcher on "how" and "why" things happen in the social world (Cooper and Schindler 2014). While qualitative techniques focus on extracting the feelings, emotions, and perceptions of study participants, they are considered fundamentally weak as their results cannot be generalized against a larger population (Cooper and Schindler 2014).

Research designs support the chosen methodology by providing a framework the researcher can use to methodically follow to address the research questions. Several quantitative research designs (experimental, quasi-experimental, and correlational) are available for use within the context of this research study. Upon evaluating the various research designs, the researcher determined the experimental research design was not appropriate for this study. Campbell and Stanley (1966) posited the experimental research design refers to investigations that involve manipulating the independent variables in the study and observing the resulting effects of that manipulation. By using the experimental research design, the researcher selects their study participants and separates them into various groups (the experimental group and the control group) to conduct their experiments on and observe the outcomes.

As part of the experimental design, the researcher has complete control of the research environment and is able to determine if there is a cause-and-effect relationship between the study variables (Campbell and Stanley 1966). In this study, the environment was not controlled by the researcher, and study participants were not categorized into experimental or control groups to accomplish the intent of the study. Additionally, Campbell and Stanley (1966) further asserted that the experimental research design may incorporate the use of multiple dependent variables, while this study only made use of a single dependent variable, further substantiating the experimental research design as not appropriate for this study.

An examination of the quasi-experimental design revealed it was also not appropriate to use within the context of this study. The quasi-experimental design (sometimes referred to as casual-comparative) involves the use of derived groups to explore the differences between two or more groups or variables (Schenker and Rumrill 2004). According to Shadish, Cook, and Campbell (2001), the quasi-experimental design shares many of the same characteristics as the experimental design, and although there is still manipulation of the independent variable(s), randomization is not present. Additionally, because there are still use of multiple groups, the quasi-experimental design is not appropriate as this study will only use a single group.

The most appropriate research design for this study was the correlational design. According to Cohen (1988), correlational designs are appropriate in studies where a quantitative dependent variable is evaluated to determine the relationship it has to one or more independent variables of interest in the study. This study used the dependent variable adoption decision and evaluated its relationship to the independent variables of this study: security effectiveness, cost savings, and reliability factor. The nature of the correlational study does not limit the form of the relationship; it can be a straight line or curvilinear (Cohen 1988).

The instrument used for this study to collect data was a modified version of the self-administered survey as adapted from Ali, Soar, Yong, and Tao (2016). The survey instrument utilized a 5-point Likert-type scale consisting of close-ended questions targeted toward IT decision-makers in government agencies, for-profit corporate organizations, and academic institutions. Data was collected using several collection methods to include online surveys, use of the survey distribution service Centiment LLC, and distribution of the survey link on social media sites like Facebook and LinkedIn. Completed survey responses were used for later correlational analysis.

Correlational research can use one of several statistical techniques to measure correlations in data. The technique this study mainly used was the Pearson correlation coefficient (r). According to Cohen (1988), the Pearson r correlation is the most commonly used technique to depict the relationship between two variables. When the r coefficient is used, it measures the degree of a linear relationship, the slope of the best-fitting straight line, and assumes normality and homoscedasticity when testing for significance (Cohen, 1988).

The data analysis commenced upon a successful collection of all survey responses related to this study. Data was entered into IBM SPSS for calculation of descriptive statistics, the Pearson correlation coefficient, and multiple correlation analysis using the continuous variables.

Research Questions/Hypothesis

IT decision-makers face a dilemma when considering whether to adopt cloud computing. The rationale for this research study was to provide IT decision-makers within various institutions additional insights to aid in the decision-making process to adopt cloud computing. The variables used to form the foundation of the research questions were the decision to adopt (dependent variable) and security effectiveness, cost saving, and reliability (independent variables). The specific research questions and hypotheses that guided this study to discover the relationship between the key factors that IT decision-makers focus on when they make their decision to adopt cloud-computing technology are the following:

RQ1: Is there a correlation between the security effectiveness factor and the decision to adopt cloud computing in a government organization, for-profit business organization, and an academic institution?

H_{01a}: There is no correlation between the security effectiveness factor and the decision to adopt cloud computing in a government organization.

H_{1a}: There is a correlation between the security effectiveness factor and the decision to adopt cloud computing in a government organization.

H_{01b}: There is no correlation between the security effectiveness factor and the decision to adopt cloud computing in a for-profit corporate organization.

H_{1b}: There is a correlation between the security effectiveness factor and the decision to adopt cloud computing in a for-profit corporate organization.

H_{01c}: There is no correlation between the security effectiveness factor and the decision to adopt cloud computing in an academic institution.

H_{1c}: There is a correlation between the security effectiveness factor and the decision to adopt cloud computing in an academic institution.

RQ2: Is there a correlation between the cost-savings factor and the decision to adopt cloud computing in a government organization, for-profit business organization, and an academic institution?

H_{02a}: There is no correlation between the cost-savings factor and the decision to adopt cloud computing in a government organization.

H_{2a}: There is a correlation between the cost-savings factor and the decision to adopt cloud computing in a government organization.

H_{02b}: There is no correlation between the cost-savings factor and the decision to adopt cloud computing in a for-profit corporate organization.

H_{2b}: There is a correlation between the cost-savings factor and the decision to adopt cloud computing in a for-profit corporate organization.

H_{02c}: There is no correlation between the cost-savings factor and the decision to adopt cloud computing in an academic institution.

H_{2c}: There is a correlation between the cost-savings factor and the decision to adopt cloud computing in an academic institution.

RQ3: Is there a correlation between the reliability factor and the decision to adopt cloud computing in a government organization, for-profit business organization, and an academic institution?

H_{03a}: There is no correlation between the reliability factor and the decision to adopt cloud computing in a government organization.

H_{3a}: There is a correlation between the reliability factor and the decision to adopt cloud computing in a government organization.

H_{03b}: There is no correlation between the reliability factor and the decision to adopt cloud computing in a for-profit corporate organization.

H_{3b}: There is a correlation between the reliability factor and the decision to adopt cloud computing in a for-profit corporate organization.

H_{03c}: There is no correlation between the reliability factor and the decision to adopt cloud computing in an academic institution.

H_{3c}: There is a correlation between the reliability factor and the decision to adopt cloud computing in an academic institution.

RQ4: What is the difference between organizations on the decision to adopt cloud computing?

H_{4}: There is a difference between adoption decisions between a government organization, a for-profit corporate organization, and an academic institution.

H_{04}: There is no difference between adoption decisions between a government organization, a for-profit corporate organization, and an academic institution.

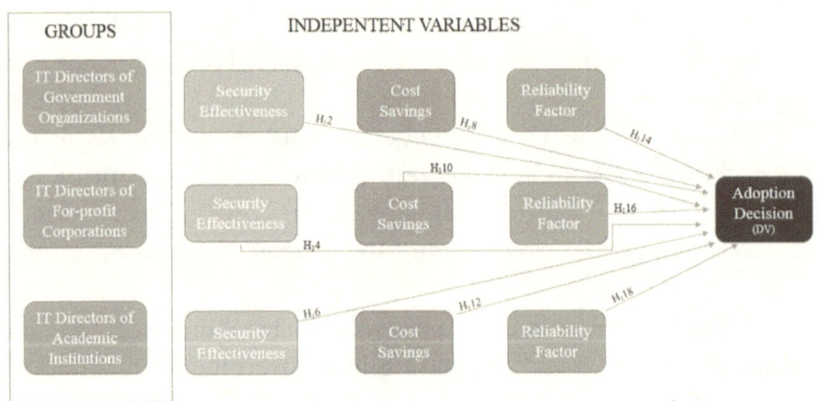

Figure 1. Operational Model

Theoretical Framework

The purpose of this study was to discover the relationship between the key factors that influence an IT director's decision to adopt cloud-computing technology within their respective institutions. The research can help provide salient information that can inform decision-makers of alternative approaches that can be utilized to increase acceptance of cloud computing within their respective organizations.

The theoretical framework utilized for this study was the Technology Acceptance Model (TAM) derived from the Theory of Reasoned Action (TRA) as discussed by Fishbein and Ajzen (1975). According to Jiang, Chen, and Lai (2010), TRA asserts behavioral intentions are determined by an individual's attitude toward the behavior and subjective norms. While according to Kim (2012), the TAM is recognized as a reliable metric and one of the predominant methods for exploring why employees accept or reject new technologies. The fundamental concept that underpins the TAM focuses on the willingness of users to accept new technologies, resulting in users altering their behaviors to adjust to the new environment (Davis 1989).

IT directors are in a critical position as they serve as the focal point between the IT department, the business operations, and the organization's customers. Decisions made regarding technology innovation of the organization's information systems have a significant impact on the business as such decisions must be approached with due diligence and caution (Lingens, Winterhalter, Krieg, and Gassmann 2016). As new technologies are realized, there are a variety of factors that affect the decision-making process as to how the technology will be used and implemented throughout the organization. According to Lingens et al. (2016), how the organization understands and accepts the technical innovation is critical to technology adoption. Several theories and models of technology acceptance have been developed over the years to uncover how users perceive and accept new technologies.

Technology has become an omnipresent force in our lives. As technology continues to evolve, researchers must continue to uncover the factors that influence technology's acceptance. Over the years, researchers have studied, developed, and proposed a variety of theories and models to explain human behavior toward technology. In this section, three technology acceptance models/theories are presented: the Technology Acceptance Model (TAM), the Theory of Reasoned Action (TRA), and the Technology Organization Environment (TOE).

The TAM is a theoretical framework used to measure the perception factors influencing a user's decision to adopt or reject the use of new technology (Jones 2009). The seminal theorist responsible for the development of the TAM was Davis (1986, 1989), who, in his seminal articles, used the constructs of perceived usefulness and perceived ease of use, which Davis found to be reliable predictors of computer use. According to Jones (2009), the TAM has a validated history of predicting user acceptance of the technology. Additionally, Stanciu (2017) stated the TAM is one of the most important approaches used in the study of people's utilization of technology, which was a revalidation of McConnell's (2009) idea that the TAM is one of the most influential and recognized models used in the field of information technology as it relates to predicting or analyzing technology acceptance. Figure 2 shows the TAM.

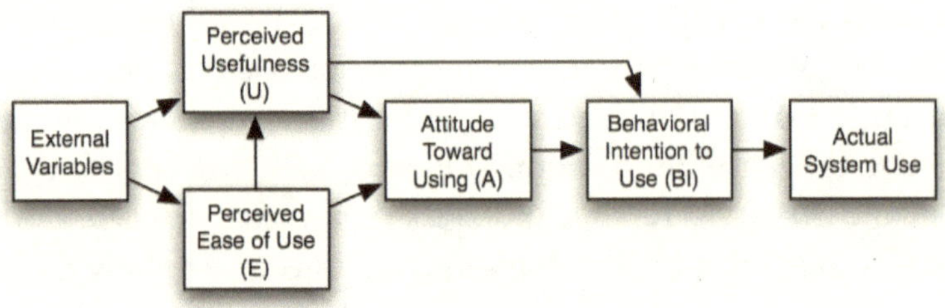

Figure 2. TAM Model, adapted from Davis and Venkatesh, 1996, p. 20

The TRA is an intention model developed by Fishbein and Ajzen (1975) rooted in behavioral theory to understand the relationships between a user's attitude, intentions, and behavior to predict a user's intention to use technology (Buabeng-Andoh 2018). The theory accepts that individual motivational factors are determinants of the likelihood of behavioral intention as influenced by a person's attitude. The behavioral intention, according to Chen, Chen, and Kinshuk (2009), is defined as an individual's inclination to recognize specific behaviors. The TRA has been used in a variety of disciplines to include marketing, health science, and technology. Figure 3 shows the TRA.

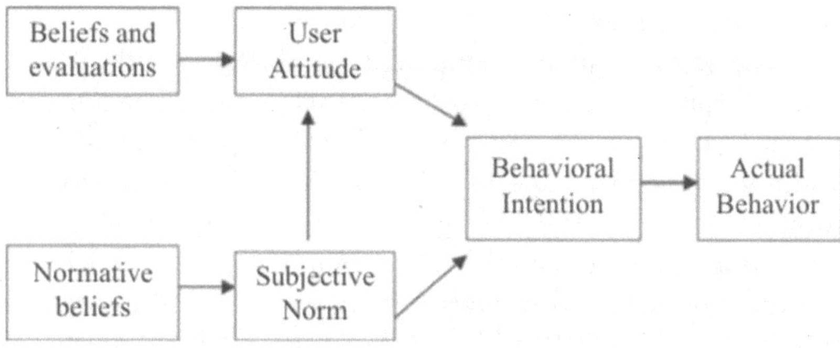

Figure 3. TRA Model, adapted from Davis et al., 1989

Unlike the TAM and TRA models that focus on individual human behavior, the TOE explores the adoption of technology from the organization's perspective (Chandra and Kumar 2018). The TOE is a theoretical framework created by Tornatzky, Fleischer, and Chakrabarti (1990) that discusses the factors that influence the likelihood of an organization adopting new technologies (Tornatzky et al. 1990). According to Chandra and Kumar (2018) and Tornatzky et al. (1990), from the organization's perspective, the decision to adopt is not contingent upon the technology alone; the decision process also considers influences of three contextual categories that include technological, organizational, and environmental contexts.

The technological context considers the impact the technology can have on the organization's enterprise systems, while the organizational context examines the impact on characteristics of the organization, which includes resources, internal communication, peer influence, and organizational culture (Qiong, Yue, and Barnes 2017). The external context is driven by government policy, competition in the marketplace, and market trends, factors that are outside the control of the organization (Tornatzky et al. 1990).

The TOE framework has been extensively used in previous research by Oliveira and Martins (2011) to analyze the assimilation of various technologies within an IT organization. Additionally, Gangwar, Date, and Ramaswamy (2015) synthesized the TAM with the TOE framework to determine the factors that directly influenced the decision to adopt cloud-computing technology.

Literature Review Strategy

The research strategy employed for this study utilized relevant and applicable scholarly/peer-reviewed journals. Literature was related to cloud, cloud computing, cloud computing in academic institutions, cloud computing in education, cloud computing in government, factors influencing cloud-computing adoption, cloud computing in business, and cloud-computing business models. The search for primary and secondary sources was limited to literature published between 2010 and 2019 to retrieve relevant, salient, and empirical material. The databases included EBESCO-host, ProQuest (ABI/INFORM), ResearchGate, SAGE Journals, Directory of Open Access Journals (DOAJ), and Emerald. The search terms used for this study included: *cloud computing, cloud technologies, cloud adoption, cloud policy, public sector, security, reliability, access, technology acceptance model, theory of reasoned action, technology organization environment, quantitative frameworks, quantitative survey instruments, quantitative data collection, validity of quantitative data*, and *reliability of quantitative data.*

Assumptions, Limitations, and Delimitations

According to Leedy and Ormrod (2016), assumptions are areas within the study that are taken for granted. In this research study, it was assumed the respondents fully comprehended the survey questions and answered all the questions honestly. Failure to complete all the questions would compromise external validity. It was further assumed the study participants kept their answers confidential from other potential study participants so as not to skew the survey results. Another assumption is all data analysis protocols were performed correctly and ethically. Finally, it was assumed each organization only had one IT director who met the criteria to participate in the study.

The limitations of this study were the sampled population was limited to participants from either a government agency, a for-profit business organization, or an academic institution of higher learning. Another potential limitation of this study was generalizing the results to the adoption of other IT-related technologies. Finally, this study only focused on participants from the Central Florida region of the US.

The scope of this study examined a general population of government, for-profit business, and academic organizations IT directors or equivalent position in Central Florida. The purposive sampling method was used to determine the sample population that participated in this research study. Study participants were restricted to those who filled the role of an IT director, CIO, deputy CIO, CTO, deputy CTO, or equivalent-level position within the organization and had the authority to make strategic decisions regarding the organization's IT strategy. The survey instrument was designed to accommodate participants from a variety of industries that have an IT or innovation department.

In this study, participants were delimited to IT directors, CIOs, deputy CIOs, CTOs, deputy CTOs, or equivalents who had strategic decision-making authority within their respective organizations. While the survey instrument can accommodate participants from a variety of organizational environments, it is delimited to government agencies, for-profit corporate organizations, and academic institutions of higher learning. The study was delimited to a real-world setting and the Internet where participants completed the survey instrument via the SurveyMonkey online survey service. In reference to data collection, the study was delimited based on responses from survey participants as part of the sample population.

Chapter Summary

Cloud computing has become one of the foremost technology strategies business, government, and educational organizations consider to extend their infrastructure, enhance their growth opportunities, gain new capabilities, and launch new innovations without having to procure additional hardware or software. The introduction of cloud technologies has been the driving force behind the creation of numerous jobs across various business units of an organization, which include marketing, sales, customer service, and accounting and finance (Tripathi and Jigeesh 2013). For many organizations, cloud technologies have become the catalyst for revenue and an enabler for new business models. Cloud computing has altered the nature of the relationship between an organization and its customers. In addition to the benefits and advantages, cloud computing has made in traditional for-profit business organizations, cloud technologies offer government agencies solutions for automating and streamlining civic processes (Daniel 2018).

Citizen services are improved through the innovation of data services in the cloud. Government agencies can reduce the footprint of their data centers by only paying for the resources needed, adding a degree of flexibility where resources can be scaled up or down to meet near-term requirements (Nikolova 2012).

In the field of academia, the technological landscape continues to change. Institutions of higher learning continue to face reduced operating budgets and are forced to adopt new technologies that enable their e-learning systems to keep pace with technology while reducing costs and staying within federal compliance regulations (Saini and Kauer 2017). The obstacles educational institutions encounter are the security of their programs and data, the challenges of moving their applications to a different service provider if the quality of service declines or the cloud service provider ceases operations, and the distribution of their data across national borders (Rajaraman 2014a).

The next chapter includes a detailed review of literature on cloud computing and the various factors that influence adoption.

2

Literature Review

The purpose of this quantitative, nonexperimental, correlational, relational survey study was to discover the relationship between the key factors that influence IT directors' decisions to adopt cloud-computing technology within their respective for-profit corporate entities, academic institutions, and government agencies in Central Florida. According to Ridley (2012), literature reviews serve to provide a historical background of the area being studied. Providing the history of a subject helps the researcher and audience understand existing literature that supports the identified research gap. Additionally, Ridley (2012) stated the contemporary context of the literature review references the current state of the topic within the field of study. As such, this literature review contains a synthesis of relevant historical and current contextual literature in the field of cloud-computing adoption.

While several theories and models have been developed for technology acceptance, this study is underpinned by the TAM originally proposed by Davis (1989). The TAM measures the perception factors that influence a user's decision to adopt or reject the use of new technology. Davis (1989) found the constructs of perceived usefulness and perceived ease of use were determined to be reliable predictors of a user's behavior toward the usefulness and ease of use of new technologies.

Title Searches and Documentation

The key words that were used to search for relevant and empirical material were the following: *cloud computing, cloud technologies, cloud adoption, technology acceptance model, the theory of reasoned action, technology organization environment, quantitative frameworks, quantitative survey instruments, quantitative data collection, the validity of quantitative data,* and *reliability of quantitative data.*

The search for primary and secondary sources was limited to literature published between 2010 and 2019 to retrieve relevant, salient, and empirical material. The databases included EBESCO-host, ProQuest (ABI/INFORM), ProQuest Digital Dissertations, ResearchGate, SAGE Journals, Directory of Open Access Journals (DOAJ), and Emerald.

Historical Content

Cloud computing has now been around for several years. Seminal research dates back to 1962, when renowned computer scientist Dr. Joseph Licklider posited the first ideas for cloud computing in a paper that discussed the intergalactic network and the man-computer symbiosis (Malley 2015). While according to Kintomide (2013), Professor John McCarthy, who was recognized as one of the founding fathers of artificial intelligence, predicted one-day computing services would take on the likeness of a utility company, in which the use of computers would be in an interactive mode. Whereby, many users would time-share the computing resources and only pay for the computing resources they used, akin to a utility company (Rajaraman 2014b). These collective ideas formed the foundation for the digital age and the concepts of what we now know as user-oriented computer networks and cloud computing. However, these concepts took more than fifty years to materialize because early research uncovered a pervasive level of confusion as to what cloud computing was (Daylami 2015).

Although Google CEO Eric Schmidt is credited with the introduction of the term *cloud computing* during an industry conference in 2006, the research uncovered a document that traced the origins of the term to two former Compaq Computer employees, George

Favaloro and Sean O'Sullivan (Regalado, 2011). In the internal document dated November 1996, O'Sullivan penned he envisaged a cloud-computing strategy for the future. The first documented use of the term *cloud computing* came in 1999, when Salesforce introduced the term *Software as a Service* and became the first company to offer the delivery of applications via the Internet (Tripathi and Jigeesh 2013). However, according to Regalado (2011), Google and Amazon were the first in the industry to use the term *cloud computing*, which they referred to as a service that incorporates the movement of applications to the Internet. By 2006, both Amazon and Google implemented their respective cloud offerings with Amazon Web Services and Google Docs, followed by Microsoft's launch of their cloud platform Azure (Tripathi and Jigeesh 2013).

Cloud computing has revolutionized the way business is conducted and how organizations interact with their customers. The seminal research into the factors contributing to the adoption of cloud computing in business identified the characteristics of security, privacy, cost savings, accessibility, reliability, scalability, and interoperability as key concerns when considering the adoption of cloud computing (Sobragi, Gastaud Maçada, and Oliveira 2014). Reese (2009) found the characteristic of security is continually mentioned as an important factor when considering the adoption of cloud computing, which includes the related activities of disaster recovery, control of the data, segmentation of the data, and network security, while Laudon and Laudon (2003) suggested security refers to the policies and procedures implemented to prevent unauthorized access to an organization's information system. Regardless of definition, all researchers agreed security is at the top of the concerns as it relates to the adoption of cloud computing in business. These findings were further supported by Tripathi and Jigeesh (2013), who opined that security is the critical component influencing the adoption of cloud computing, and according to Gonzalez and Smith (2014), security was one of the considerations causing small- to medium-sized companies to reject adopting cloud computing.

Early research also showed cost savings was another common factor that contributed to the decision to adopt cloud computing over maintaining traditional in-house IT systems (Ojala and Tyrvainen 2011). Research conducted by Reese (2009) substantiated these findings stating the cost savings generated by the adoption of cloud computing were significant and could expect to increase exponentially as the underlying infrastructure matures and can handle additional capacity. However, according to Misra and Mondal (2011), while most research finds the implementation of cloud computing is accompanied by cost savings, some groups disagree and posits the initial costs involved in moving to a cloud-based service model do not provide an acceptable return on investment.

Research into cloud computing within the field of academia has revealed a growing trend in the adoption of cloud-based services. Georgescu and Matei (2013) found academic institutions gain benefits from implementing cloud computing by being able to shift resources against the organization's priorities as opposed to investing time and money, maintaining the university's IT infrastructure. In this context, academic institutions can provide education services more efficiently at a lower cost (Georgescu and Matei 2013). These concepts are further supported by Sultan (2010), who posited cloud computing as an attractive option for educational institutions seeking to take advantage of new technology developments with moderate out-of-pocket expenditure.

Researchers in the field agree that the value academic institutions can reap include the elimination of up-front costs for hardware, software, training, and operations and maintenance (Katzan 2010). Additionally, opportunities to leverage economies of scale are realized by moving education software and services to a cloud environment. As academic institutions seek new ways to keep up with the ever-changing technological landscape, having their software available in the cloud helps to attract new students with the variety of productivity tools available (Burfield 2011, Sultan 2010), resulting in increased collaboration, communication, and opportunities for distance learning (Burfield 2011).

In today's technology-aware markets, organizations across all industries need to innovate to be competitive. Organizations constantly seek to find new innovations that can be implemented across the business while finding ways to lower costs for the organization, and the government is no different. Government agencies are under significant pressure to improve the speed at which they implement new citizen services, drive new efficiencies, and reduce capital expenditures (Roster, Moore, and Pfeiler 2010). The seminal research into cloud-computing adoption within the government sector suggested the newness and underdevelopment of the technology as significant challenges to adoption; however, according to Nikolova (2012), adoption decisions were driven by both technical and financial factors. Paquette, Jaeger, and Wilson (2010), on the other hand, believed the adoption of cloud computing in government agencies was hampered by significant tangible (unauthorized access, service provider equipment failure, and service availability) and intangible risks (user confidence in the capability and public access). The approach the government would take to manage and mitigate these risks, however, would be key factors in the adoption of cloud computing (Paquette, Jaeger, and Wilson 2010).

While cloud computing is less popular in government agencies, it has the potential to integrate the elements of lowering the cost for services, improved flexibility, increased IT agility, and improved collaboration throughout the organization (Nikolova 2012, Oberer and Erkollar 2012). Oberer and Erkollar (2012) further supported this theory as they opined governments needed to adopt cloud technology, citing the complexity of future economic constraints on financial budgets made available to government agencies. Iovan and Daian (2013) also agreed that governments must look to adopt innovations like cloud technology to reduce the financial burdens citizens are obligated to pay toward their states in the form of taxes and service fees. Before deciding to adopt cloud computing, Schultz (2011) suggested governments evaluate specific applications for transitioning to the cloud in addition to weighing the factors of security, legality, and compliance. Research conducted by numerous scholars found security to be a common factor to be considered in the decision to adopt cloud computing. However, Oberer and Erkollar (2012) and Solomon (2009) concluded the legal framework, privacy, performance, availability, and ability to customize as key factors that have historically hampered cloud adoption in government markets.

Current Content

Previous research into the adoption of cloud computing between business organizations, government agencies, and academic institutions revealed several benefits of implementing cloud-computing technology. Among the factors, these organizations considered as significant in the decision-making process were security, privacy, and cost (Gonzalez and Smith 2014; Misra and Mondal 2011; Sobragi, Gastaud, Maçada, and Oliveira 2014; Tripathi and Jigeesh 2013). The next section of this paper will review and synthesize the current literature as it pertains to the adoption of cloud technologies among business organizations, government agencies, and academic institutions.

While a plethora of research has been conducted on cloud computing with evidence of benefits in all industries Ali, Soar, Yong, and Biswas (2015); Buyya, Yeo, and Venugopal (2008); and Marston, Li, Bandyopadhyay, Zhang, and Ghalsasi (2011) all agree there is a dearth of research that provides a comprehensive investigation into the factors that should be considered as organizations seek to adopt cloud computing. Additionally, according to Hassan, Nasir, Khairudin, and Adon (2017), extant research into the factors that influence the adoption of cloud computing is lacking. However, as cloud computing has shown to offer businesses several benefits, companies must strategically position themselves to adopt new innovations that enable them to alchemize their business processes.

Although some in academia have opined that research into the factors that influence cloud computing adoption is lacking, Habjan and Pucihar (2017) found 22 percent of European Union countries enterprises have already adopted cloud computing, with the components of service value and service connectivity representing the most important factors influencing the decision to adopt. Contrasting this idea, Lian, Yen, and Wang (2014), in their research, found the most important factors to consider when deciding on the adoption of cloud computing were the factors of data security, perceived technical competence, and cost.

While early research focused on the aspects of the system (i.e., security, privacy, performance, etc.), current research analyzed factors external to the system. Raza, Adenola, Nafarieh, and Robertson (2015) postulated the lack of regulation within the business industry has significantly contributed to the slow adoption of cloud computing. Additionally, the introduction of globalization and big data has posed new challenges to organizations that forced businesses to restructure their business models and how they deliver IT services via a cloud-computing model (Raza et al. 2015, Shawish and Salama 2014).

In today's technologically advanced world, institutions of higher learning have become reliant upon technology for content delivery, communication, and increased collaboration. Bhatia (2014) stated that upward of 90 percent of organizations currently use cloud computing or some derivative of within their daily operations. According to Bernsteiner and Pecina (2015) and Pardeshi (2014), 28 percent of US institutions of higher learning have already adopted and implemented cloud-computing technologies, while another 29 percent of institutions had cloud technologies on their technology road map for future adoption.

Academic institutions gain immediate benefits through cloud computing's dynamic scalability and effective use of resources. According to Saini and Kaur (2017), the ability for cloud computing to reduce IT costs remains a top factor in the decision to adopt cloud-computing technology within the education sector; however, its ability to create modern collaborative environments increase opportunities for academic institutions to extend learning beyond the traditional classroom. Muniasamy, Ejalani, and Anandhavalli (2014) supported this premise in their research as they suggested computer usage around the world is low as a result of the high cost of computer hardware and maintenance; however, adopting cloud-computing technology is the optimal solution for meeting the educational dilemma. Additionally, as the number of students and academic content continues to increase, institutions of higher learning are faced with escalating costs and shrinking budgets. Muniasamy, Ejalani, and Anandhavalli (2014) posited these elements as significant factors in academic institutions seeking to adopt cloud computing in addition to its flexibility, scalability, speed of implementation, and ubiquitous access.

The speed at which technology is advancing has introduced additional factors to be considered when adopting cloud computing. According to Gangwar, Date, and Ramaswamy (2015), the phenomena of globalization has impacted the way academic institutions operate and has essentially forced them to adopt innovative technologies like cloud computing to cut costs while improving their operational efficiency. Additional factors considered in the decision to adopt cloud computing include technology compatibility, complexity, data security, compliance with regulatory standards, and service reliability (Njenga, Garg, Bhardwaj, Prakash, and Bawa 2019). Several predictor factors have also been cited as affecting the decision to adopt cloud computing, technology factors encompassing the relevance, usefulness, and ease of use of the technology. Organizational factors refer to the organization's readiness to adopt cloud computing, the level of management support for the adoption, and acknowledgment of the perceived barriers to adoption. Finally, the environmental factors stemming from competition within the industry (Low, Chen, and Wu 2011; Njenga, Garg, Bhardwaj, Prakash, and Bawa 2019).

As more academic institutions adopt cloud technologies and move their data to cloud-based infrastructures, the need for trained IT staff with expertise in cloud implementations has grown. During a recent survey, Gonzalez et al. (2012) found the top three major security issues impacting cloud-computing adoption in educational institutions were legal issues, compliance, and loss of data control. To aid in mitigating these factors, Attaran, Attaran, and Celik (2017) identified a three-phase strategy for successfully adopting and implementing cloud computing in institutions of higher learning. The implementation of an elementary phase where the institution performs an in-depth evaluation of their internal applications and the services they will require was done. As part of its analysis in phase two, the institution would evaluate how cloud implementation would impact their internal processes. In the final phase, internal school applications are mapped to a corresponding cloud service (Attaran, Attaran, and Celik 2017).

In the years since the paradigm shift to cloud computing, the government has taken major steps in adopting cloud-computing technology by issuing a "cloud-first" mandate, whereby government agencies were directed to take strides toward implementing some variant of cloud computing (US Agencies Embrace the Cloud 2014). Unlike other governmental agencies, the military has special organizational and technical requirements that need to be considered when deciding to adopt cloud-computing technology. According to Ďulík and Ďulík (2016), the following factors have been found important from the perspective of the Department of Defense as part of the decision-making process for adopting cloud computing: cost of entry, data location, security, compliance, and application programming models.

In contrast to their federal counterparts, the current research found state and local government agencies have migrated to cloud implementations exponentially faster as a result of having less restrictive factors they consider when deciding to adopt cloud computing. According to Daniel (2018), a recent survey found 80 percent of the state and local government agencies had already adopted some variation of cloud computing, with 75 percent of those states observing increased adaptability and 38 percent reporting observable cost savings. The top factors influencing the decision to adopt cloud-computing technology among these government leaders were cost, scalability, and compliance with regulatory requirements (Daniel 2018).

Through all the historical and current research on cloud adoption, Ali, Soar, Yong, and Tao (2016) found data security and privacy topped the list of factors influencing the adoption of cloud computing in the government sector. Despite the vast literature that exists on cloud computing, academia and industry professional's conflict as to an agreed-upon definition for what cloud computing is exists. While there is no universal definition for cloud computing, several definitions from the industry and academic perspectives are discussed below.

According to Benlian, Kettinger, Sunyaev, Winkler, and guest editors (2018), cloud computing is defined as a service provided by computing vendors that enable ubiquitous on-demand access to a collection of shared configurable computing resources, a technology that enables individuals and organizations to access services from a myriad of platforms as a measured service. While according to research conducted by Madhavaiah, Bashir, and Shafi (2012), the 451 Group defined cloud computing in terms of combining the aspects of grid computing, virtualization, and Software as a Service into a deliverable service model. While AMR research suggested cloud computing is an environment whereby consumers pay a subscription for software services and rent IT infrastructure as their needs demand. The Enterprise Strategy Group posits cloud computing as an alternative service model for purchasing IT equipment. Instead, the business purchases IT performance, availability, data protection, and security as a suite of services the business relies on to execute their mission (Madhavaiah et al. 2012). Cloud computing was defined by the Gartner Group as a service delivery method that provides scalable IT capabilities to the end user.

Academic research conducted by Briscoe and Marinos (2009) defined cloud computing as the use of Internet-based technologies to provision resources that provide scalable services over the Internet. Buyya, Yeo, and Venugopal (2008) compared cloud computing to a form of a distributed computing environment that uses a collection of provisioned virtualized computers and presented them as a scalable resource to customers. Similarly, Foster et al. (2008) described cloud computing as a distributed computing paradigm that leverages interconnected virtual computers to deliver managed scalable computing power, storage, and platform as an on-demand service. In the view of Marks and Lozano (2010), cloud computing is defined as a service model that provides on-demand access to virtualized computing resources over the Internet via a paid subscription. According to Grossman and Gu (2009), cloud computing provides infrastructure services via the internet. In the opinion of Nurmi et al. (2008), cloud computing consists of dynamically provisioned services that range from Infrastructure as a Service and Software as a Service that is provided on demand. In all the definitions of cloud computing, a common theme is identified. Cloud computing is a new paradigm by which shared hardware and software resources are provisioned on demand over the Internet.

Early cloud-computing research in business showed the technology to be as innovative and disruptive as the diffusion of telecommunications infrastructure was in the 1970s (Madhavaiah, Bashir, and Shafi 2012). Organizations found through the use of cloud computing they were able to rent computing resources using an on-demand format as they already did for energy and electric services, which resulted in having an overwhelming impact in the following areas (Etro 2011):

- Establishment of new business
- Improved national economy
- New jobs across various industries
- A redistribution of job in the IT sector

The underlying technologies supporting cloud implementation allow users and organizations to outsource their computing needs utilizing an on-demand service model consisting of three categories: Infrastructure as a Service (IaaS), Platform as a Service (PaaS), and Software as a Service (SaaS) (Kim 2017). According to Goncalves and Ballon (2011), the principal rationale for the use of cloud computing among organizations is to reduce internal IT costs by procuring IT-related capabilities from a third-party service provider. Understanding the various service offerings available is important when determining a suitable service model. Virtualization technology is the underlying foundation that provides the primary characteristics of cloud computing, namely location independence, resource pooling, and scalable provisioning (Kim 2017).

IaaS is a service model that provides an organization the capability to provision compute, storage, memory, networks, and various other IT-related resources where the customer can deploy arbitrary software (Kim 2017). In this service model, the organization outsources its hardware and software requirements, and services are charged by usage and can be scaled dynamically as an on-demand service. The underlying physical infrastructure is managed by the service provider; however, the operating systems, storage, and applications are controlled by the organization (Gutierrez-Garcia and Sim 2013). According to Akintomide (2013), Amazon's Amazon Web

Services and Amazon EC2 (Elastic Cloud Computing) are prime examples of services offered under the IaaS model.

One of the hidden dangers of IaaS is the simplicity of getting started with it. According to Gibbons (2019), users across the organization can take advantage of services such as deploying new virtual machines, SQL databases, monitor streaming video, and create backup servers with no organizational oversight, resulting in negative impacts to departmental budgets.

The implementation of a Platform as a Service (PaaS) provides the ability for an organization to deploy infrastructure in the cloud environment, thereby providing access to capabilities required to develop and operate applications via the Internet (Akintomide 2013). The seminal research into PaaS revealed software development teams were the primary users of this service model as it provided an elastic on-demand environment where the use of programming languages, libraries, services, and tools was supported by the cloud service provider (Krancher, Luther, and Jost 2018; Rittinghouse and Ransome 2016). PaaS has shown considerable growth in popularity within the industry as according to research conducted by Gartner (2018), the demand signal for PaaS offerings grew from $3.8 billion in 2015 to $7.2 billion in 2016 with expected growth rates of 20 percent per year. Similar to IaaS, the customer does not control the underlying cloud infrastructure; however, the customer has full control over deployed applications hosted in the environment (Rittinghouse and Ransome 2016). PaaS as a service offering has proven to more than double developer productivity. According to research conducted by Cloud Foundry (2012), 46 percent of survey respondents reported shorter development cycles over one week after adopting PaaS.

According to Kraft (2018), the SaaS model hosts applications on cloud infrastructure and renders them available to customers on a subscription basis. Customers have no control over the underlying infrastructure; however, they have full control over the applications being hosted. Using this service model, an organization can reduce its IT cost by having the vendor host the applications while the organization only pays for the licenses they need (Kraft 2018). Early adopters of SaaS were risk averse and only considered moving noncritical system functions to the cloud environment. However, according to Heart (2010), now that the technology has been in the marketplace for a while, organizations are willing to take chances and move their core IT functions and services to SaaS-based platforms. Organizations have realized the financial benefits of moving their IT services to a SaaS subscription model. According to Kraft (2018), the purchase of software licenses, renewal of annual maintenance costs, and IT infrastructure are no longer required as they are furnished by the SaaS service provider as part of the subscription service.

While SaaS includes several benefits, Kraft (2018) posits there are several disadvantages associated with the implementation of SaaS. Among the top concerns are security, the ability to integrate with third-party applications, and the nuances of changing service providers (Kraft 2018).

Theoretical Framework Literature

Theoretical frameworks provide the foundation and context upon which the research findings are grounded. Several theoretical frameworks in the adoption of technology acceptance have been identified as having demonstrated robustness in studies that focus on the adoption of new and emerging technologies (Koul and Eydgahi 2017). Present research in the field of technology adoption relies on a composite of the three most favored models used for predicting user acceptance of new technologies: the Technology Acceptance Model (TAM), the Theory of Planned Behavior (TPB), and the Diffusion of Innovation theory (DOI). Additionally, Taherdoost (2018) postulated several theories that are used to explain the relationship between attitudes and behavior, which also influence the decision to adopt new technologies. These theories are the Theory of Reasoned Action (TRA), Model of PC Utilization (MPCU), Social Cognitive Theory (SCT), and the Motivational Model.

According to Koul and Eydgahi (2017), the main purpose of the TAM is to predict user acceptance of new technologies and to draw attention to potential design flaws within the system before mass deployment. The TAM uses the constructs of perceived usefulness (PU) and perceived ease of use (PEOU) as the perceptions users have regarding the new technology or system (Dillon and Morris 1996). In the context of the research study, PU is described as the degree a user feels the use of a system would improve their job performance, while PEOU is the degree of a user's belief that the referenced systems will be effortless to use (Davis 1989). Together, these constructs help establish a framework for explaining a user's behavioral intention to adopt new technology (Koul and Eydgahi 2017). Although the TAM is considered one of the most dominant and effective frameworks for technology acceptance, it has gone through several modifications or extensions, which include TAM2 and unified theory of acceptance and use of technology (UTAUT) (Koul and Eydgahi 2017).

TAM2 is an extension to the original TAM, which was introduced by Davis (1989). Under the TAM2 framework, Venkatesh and Davis (2000) proposed removing the component of attitude toward use while introducing the elements of social influences and a cognitive process to the model by adding the variables subjective norm, image, voluntariness, and experience as a mechanism for measuring a user's intention to use new technologies. While the cognitive process captured characteristics that were relevant in the performance of the job, output quality, and perceptible results (Lee, Li, Yen, and Huang 2010). According to Legris, Ingham, and Collerette (2003), adding the additional variables enabled the modified TAM to provide stronger, predictive, and higher explanation abilities.

The TAM and TAM2 models provided empirical foundations on the impacts of external variables to explain and predict the use and acceptance of new technologies. Although research demonstrated the TAM2 model accounted for 52 percent of the variance in usage intentions and 60 percent in usefulness perceptions, further model integration was required to target populations resistant to adopting new technology (Koul and Eydgahi 2017). To reach these target populations, Venkatesh, Morris, Davis, and Davis (2003) amalgamated several of the existing theories into a new theory termed the unified theory of acceptance and use of technology (UTAUT). Researchers found using an adjusted R2 value the UTAUT model better explained individual technology acceptance and use decisions within organizations (Koul and Eydgahi 2017, Venkatesh, Thong, and Xin 2016).

The UTAUT considers four constructs and four moderators to aid in predicting behavioral intention: performance expectancy, effort expectancy, social influence, and facilitating conditions and age, gender, experience, and voluntariness (Venkatesh, Thong, and Xin 2016). According to Venkatesh, Thong, and Xin (2016), the constructs of performance expectancy, effort expectancy, and social influence were found to affect behavioral intention to use technology while behavioral intention coupled with facilitating conditions determine technology use. Weber (2012) posits evaluation of the UTAUT model found it to be a high-quality theory that clearly defines and articulates its various parts. Additionally, because the UTAUT synthesizes several existing theories, it performs well in focusing on the phenomena of technology acceptance and use while introducing high-order moderation effects and rigorous empirical validation (Venkatesh, Thong, and Xin 2016).

The TPB is foundationally based on the Theory of Reasoned Action introduced by Ajzen and Fishbien (1980). According to Koul and Eydgahi (2017), the TPB was an extension and improvement of the original theory, which mainly focused on the prediction of planned human behavior while factoring in the constructs of perceived behavioral control, subjective norms, and attitudes. According to Boslaugh (2008), TPB is the most commonly used behavioral model capable of explaining and predicting human intention.

Perceived behavioral control is the construct that asserts some beliefs account for both internal and external factors that can influence the performance of a behavior (Cheng et al. 2019). The construct of subjective norm refers to the beliefs that result from an individual's personal perception of what the normative expectations of others would be and the perceived social pressure to perform the behavior of interest (Cheng et al. 2019). The construct of attitude is theoretically comprised of two components: affective attitude and evaluation attitude (Cheng et al. 2019). Affective attitude refers to both the pleasure and displeasure experienced by an individual as a result of executing specific behaviors, while evaluation attitude assesses the perceived advantages or ramifications of those behaviors (Cheng et al. 2019).

Together, the three considerations of PCB, subjective norms, and attitudes made the TPB model more suitable for predicting system usage (Koul and Eydgahi 2017). The TPB model has been empirically validated and supported by the numerous studies it has been used to predict and explain social behavior (Boslaugh 2008). However, according to Chuttur (2009), when compared with the TAM, the TPB model provided additional information used to explain users' intent to use.

While numerous studies have shown the accurate predictability of the TPB model, some in academia have proposed incorporating additional components to the model to improve its predictability (Boslaugh 2008). The recommended additions include "expectation, desire, and need; affect and anticipated regret; personal or moral norm; descriptive norm; self-identity; and past behavior and habit" (Boslaugh 2008, p. 1034). Figure 4 shows the TPB.

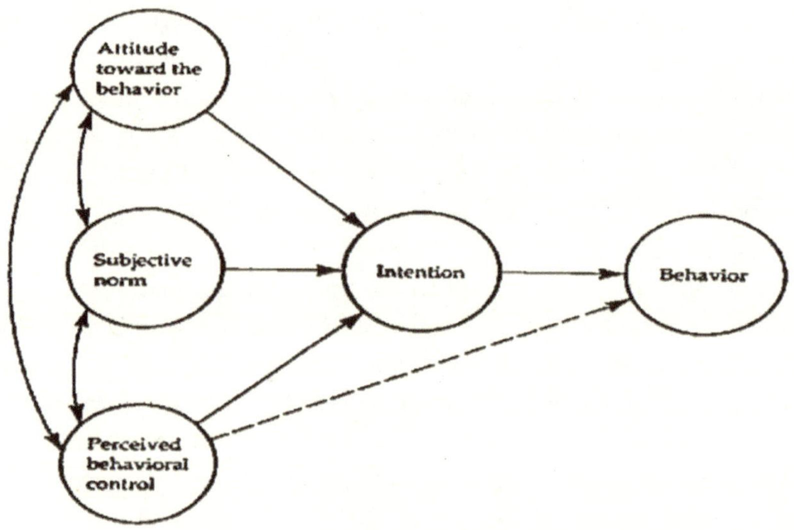

Figure 4. Theory of Planned Behavior Model, adapted from Ajzen 1991, p. 182

The diffusion of innovation (DOI) theory is based on the concept that the process for communicating an innovation within an organization is disseminated to members of the social system via specific channels over a specified period (Yapp, Balakrishna, Yeap, and Ganesan 2018). The primary objective in the DOI is to determine the degree to which an innovation spreads between the various types of adopters. The seminal theorist behind the DOI theory was Everett Rogers, who defined diffusion as a process that results in social change when new ideas are presented (Gouws and Rheede van Oudtshoorn 2011). Zanello, Fu, Mohnen, and Ventresca (2016) affirmed there are two elements involved in the diffusion process: "the innovation must have a nature suitable for the context in which it is spreading, and vectors of communication diffusion must be in place in order to transmit information" (p. 888).

Gouws and Rheede van Oudtshoorn (2011) described the DOI theory as involving four elements: the innovation itself, communication, the social system, and time. According to Gouws and Rheede van Oudtshoorn (2011), innovation is a new idea, practice, or technology to be considered for adoption. Communication, which is at the center of the diffusion process, refers to the dissemination of information regarding the new concept to others (Gouws and Rheede van Oudtshoorn 2011). The social system is the group the information is communicated to. The social system plays a large part in either the acceptance or rejection of the new innovation. The social system can either facilitate diffusion of the innovation or advocate its rejection (Gouws and Rheede van Oudtshoorn 2011). Finally, time is a critical element in the process as the longer it takes to communicate the innovation, the higher the risk the innovation will be rejected.

According to Rogers (1995), there are five groups of technology adopters, the innovators who are prone to taking risks as it relates to technology. The early adopters, a risk-tolerant group that understands technology, embraces innovation and is open to accepting new technologies. The early majority is represented as a subset of the early adopters group that advocates the technology's acceptance (Rogers 1995). The late majority, the fourth group, is represented by individuals who wait until the technology is at the peak of diffusion before accepting it. In the final group, the laggards are representative of individuals who adopt the technology after its usage has been saturated in the market (Rogers 1995).

Research into behavioral theories revealed TRA is the most commonly used theory to study the phenomena of behavioral intentions (Chipidza and Green 2019). At the center of the theory was the construct of behavioral intention, which is driven by attitudes and social norms. Fishbein and Ajzen (1975) discussed how individuals devise their intentions and how those intentions are influenced by the factors of attitudes, subjective norms, and expectations from other people. Similar to the TPB model, attitudes refer to the beliefs of executing a specific behavior and the consequences associated with those behaviors, while the subjective norm refers to those beliefs that result from an individual's perception of what the perceived social norm would be and the perceived social pressure to perform the behavior of interest (Cheng et al. 2019). The combination of attitudes and the subjective norms have empirically shown to influence behavioral intent; however, one noted deficiency was the TRA model was unable to distinguish between volitional and nonvolitional behavior (Wiley and Corey 2013). To improve the TRA model, Ajzen (1985) added the construct of perceived behavioral control to capture perceived behaviors that were not of an individual's own free will. The addition of perceived behavioral control accounted for the components of internal and external control. According to Chipidza and Green (2019), internal control referred to an individual's belief in their abilities to accomplish a specific behavior, while external controls account for those external factors that can influence their behavior (Ajzen and Madden 1986). Figure 5 shows the TRA.

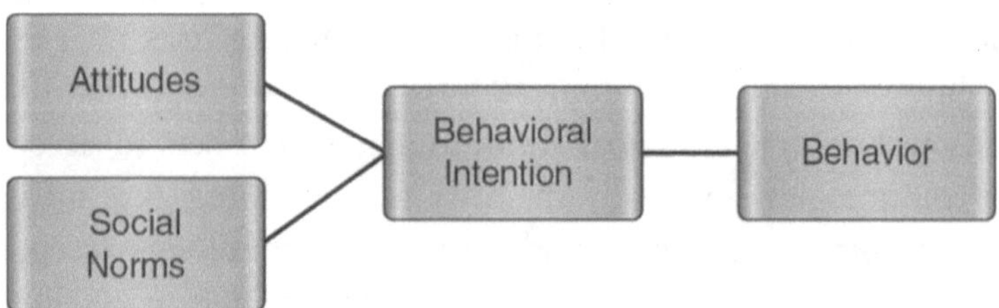

Figure 5. Theory of Reasoned Action Model, adapted from Wiley and Cory, 2013, p. 646

The MPCU was derived from Harry Triandis's (1977) theory of human behavior. The MPCU model introduced new perspectives that vastly opposed the concepts originally proposed in the TRA and TPB models (Jawahar and Harindran 2016). Triandis's human behavior model proposed the likelihood of an individual performing certain behaviors is contingent upon the presence of certain criteria, which included previous experiences performing the behavior of interest, the individuals' intent to perform the behavior in the future, and the conditions that make it easier for the behavior to become a habit (Kerr 1989).

Triandis (1977) asserts that behavioral intention is comprised of several dependent factors: an individual's assessment of their feelings about the behavior, their beliefs regarding the consequences of performing the behavior, and the subjective norms as it pertains to the appropriateness of the behavior. In addition to the perceptions of the subjective norms, behavioral intention also accounts for the perceived appropriateness of the behavior from an individual's personal normative belief (Kerr 1989, Seibold and Roper 1980).

Understanding the value human behavioral theory plays in the adoption of technology, Thompson, Higgins, and Howell (1991) suggested the adoption of Triandis's human behavior model. Using the constructs of job fit, complexity, long-term consequences, affect toward use, social factors, and facilitating conditions, Thompson et al. (1991) was able to synthesize the human behavior model with an Information Systems context to develop a new model that could be used to predict acceptance and usage of information technologies (Jawajar and Harindran 2016).

Developed from the social learning theory, the SCT postulated that behaviors are developed by observing others (Baez, Hoch, and Cramer 2019). The fundamental concept underpinning the SCT is an individual's self-efficacy greatly influences both their behavior and performance (Chung and Park 2019). While an influential theory in human behavior, Compeau and Higgins (1995) extended the theory for use in determining individual acceptance and use of information systems while incorporating the core constructs of outcome expectations performance, outcome expectations personal, self-efficacy, affect, and anxiety.

Empirical research into technology acceptance has uncovered a variety of theoretical models that have been used to explain or predict the adoption and use of technology. While some models focused on behavioral intention and others on perceived ease of use and perceived usefulness, the MM approaches technology acceptance from the motivational perspective (Sällberg and Bengtsson 2016). The MM is based on the self-determination theory as proposed by Deci and Ryan (1985), which discussed the differences between intrinsic (personal joy) and extrinsic motivations (peer pressure). Davis, Bagozzi, and Warshaw (1992) "applied motivational theory in the information systems domain to understand the factors influencing the adoption and use of new information systems" (Jawajar and Harindran 2016, p. 37).

Methodological Literature

In previous research that studied the factors influencing the adoption of cloud computing, Ross (2010) employed quantitative methods, including the use of a validated survey instrument developed by Lease (2005) to evaluate the factors that influence a business organization to adopt cloud computing. Ross (2010) approached the study from the lens of decision-making professionals, which included chief information officers (CIOs), information technology (IT), information systems (IS) managers, and other managers involved in the decision-making process who have an influence on IT policy decisions within the organization. Ross concluded there were strong relationships found between the independent variables of cost-effectiveness, the need for cloud computing, reliability, and perceived security effectiveness and the dependent variable of management interest in adopting cloud computing.

In a study to identify the key factors that influence CIOs' decision to adopt cloud computing in higher-education institutions, Alfifi (2015) evaluated the perceptions of CIOs toward the factors of security effectiveness, cost saving, data integrity, reliability, and accountability of cloud computing. To achieve the goals of this study,

Alfifi employed exploratory research utilizing a quantitative methodology. A validated survey instrument was also used in this study to collect the data. According to Alfifi, the survey instrument was based on previous research conducted by numerous researchers in academia to include Ross (2010), which was a derivative of the original survey developed by Lease (2005). Alfifi's (2015) quantitative study revealed the factors of security effectiveness, cost saving, data integrity, reliability, and accountability significantly influenced CIOs' and other key members' decision to adopt cloud computing.

Additional research by Obinkyereh (2017) focused on the factors that determine cloud-computing adoption by IT professionals in Ghana. Obinkyereh also employed a quantitative correlational survey that focused on measuring the relationship between the independent variables perceived ease of use, perceived usefulness, perceived benefit, perceived security, and perceived accessibility and the dependent variable cloud-computing adoption in Ghana. Obinkyereh grounded the study using the TAM as the theoretical framework supporting the foundation of this study. Statistical analysis found there were significant relationships between the variables perceived usefulness, perceived benefit, perceived security, and perceived accessibility and cloud-computing adoption in Ghana while the variable perceived ease of use did not have a significant relationship contributing to the decision to adopt cloud computing in Ghana.

Finally, Gonevski (2018) conducted a quantitative study to determine which factors were significant in shaping the attitudes of IS managers toward the adoption of cloud computing within their organizations. Similar to previous research in the area of cloud-computing adoption, Gonevski grounded the study using the TAM theoretical framework. However, unlike previous research in this area, Structural Equation Modeling was used to analyze the data. SEM analysis found positive correlations between the constructs perceived usefulness and perceived ease of use with adoption attitudes as well as positive correlations between perceived usefulness, perceived ease of use, and perceived benefits with Information Channels Exposure (Gonevski 2018).

Research Design Literature

Previous research conducted by Ratten (2013) explored the behaviors and perceptions that affect the decision to adopt cloud computing. Using the case study research design, Ratten (2013) found technology innovations affect a person's emotional perspective, thereby influencing their adoption behavior. Similarly, citing the lack of exploratory studies on the adoption of cloud computing, Alsanea and Wainwright (2014) used the multiple case study research design to identify the factors that impact the adoption of cloud computing from the lens of the organizational, technological, and environmental domains. The case study research design found service quality, usefulness, security, complexity, cost, organization size, IT infrastructure readiness, senior management support, feasibility, trust, organization culture, organization structure, privacy risk, direct benefits, indirect benefits, culture, external pressure, industry type, government support, and regulatory issues were among the top concerns influencing the decision to adopt cloud computing (Alsanea and Wainwright 2014).

Additional research conducted by Hakim (2017) leveraged the grounded theory design as the foundation to generate a new theoretical model to help understand the determinants that contribute to cloud-computing adoption. Finally, Effiong (2020) used the descriptive phenomenological design to explore the lived experiences of managers to discover what factors influenced their decision to adopt cloud computing.

Conclusions

Throughout the review of literature, some patterns remained consistent from the initial debut of cloud computing through the evolving maturation of the technology as we know it today. Seminal research by Reese (2009) found the factors of security and data privacy were a recurring concern and noted as an important factor for consideration in the decision to adopt cloud computing. Later studies by Schultz (2011) suggested government agencies should prioritize the factor of security while evaluating various applications for potential transition to the cloud. Researchers concluded that organizations across the different industries evaluated the various factors influencing their decisions before adopting cloud computing. Among the factors, these organizations considered as significant in the decision-making process were security, privacy, and cost (Gonzalez and Smith 2014; Misra and Mondal 2011; Sobragi, Gastaud, Maçada, and Oliveira 2014; Tripathi and Jigeesh 2013).

While security remained a factor in the decision-making process, current research conducted by Ďulík and Ďulík (2016) added the factors of cost and compliance as being significant factors influencing the decision to adopt cloud computing. These findings were further support by Daniel (2018), who found the top factors influencing the decision to adopt cloud-computing technology among government agencies were cost, scalability, and compliance with regulatory requirements.

Chapter Summary

The purpose of this correlational study was to discover the relationship between the key factors (security effectiveness, cost savings, and reliability) that influenced IT directors' decisions to adopt cloud-computing technology within their respective institutions. The literature review conducted as part of this chapter took a comprehensive analysis of previous studies from both a historical perspective and current published written works that focused on the factors that influenced the adoption of cloud computing across a variety of industries. The seminal research into the topic of cloud computing revealed the term *cloud computing* was first conceptualized in 1996 by two former Compaq Computer employees; however, the term was not officially used within the industry until 2006, when then Google CEO Eric Schmidt introduced the term at an industry conference (Regalado 2011).

As cloud computing revolutionized the way organizations conducted business and interacted with their customers' decision-makers raised key concerns over the characteristics of cloud computing that influence the decision to adopt. Reese (2009) found the factor of security was a recurring concern in the decision-making process. These findings were supported by a study conducted by Tripathi and

Jigeesh (2013), who opined that security is the critical component influencing the adoption of cloud computing while Gonzalez and Smith (2014) found security was one of the considerations causing small- to medium-sized companies to reject the adoption of cloud computing. Current research revealed the move to cloud computing has become a popular phenomenon, as Bhatia (2014) suggested upward of 90 percent of organizations have already adopted some form of cloud computing.

Although this study was grounded using the Technology Acceptance Model as the theoretical framework, a review of the Theory of Planned Behavior and the Diffusion of Innovation Theory as validated acceptance models were also conducted. According to Koul and Eydgahi (2017), the TAM is considered one of the most dominant and effective frameworks for technology acceptance, while the TPB focused more on the prediction of planned human behavior utilizing the constructs of perceived behavioral control, which, according to Koul and Eydgahi (2017), was more suited for predicting system usage.

Chapter 3 will expand upon the methodology and design utilized in this study.

3

Research Methodology

The literature review accomplished in chapter 2 analyzed previous studies from a historical perspective as well as current published written works that focused on the factors that influenced the adoption of cloud computing across a variety of industries. Business institutions invest a significant amount of their operating budget into information systems to assist their respective organizations in accomplishing their operational goals by supporting their strategic objectives while improving employee productivity and providing the organization a competitive advantage. To ensure survivability in twenty-first-century global markets, organizations must take advantage of new technologies that facilitate the development and delivery of higher-quality services. For-profit corporate organizations, academic institutions, and government agencies have traditionally been burdened with maintaining the rising costs of their IT infrastructure and data centers that support networking equipment, servers, and storage arrays (English 2018).

As cloud computing revolutionized the way organizations conducted business and interacted with their customers, decision-makers raised concerns over the characteristics of cloud computing that influence the decision to adopt. Research conducted by Alfifi (2015) evaluated predetermined criteria of CIOs of higher-education institutions in relation to their perceptions of the security effectiveness, cost saving, reliability, and accountability of cloud computing, while Hakim (2017) conducted research with CIOs of technology- and nontechnology-oriented companies that focused on the factors that contributed to the resistance of the adoption of cloud computing.

Previous research conducted by Nikolova (2012) revealed the use of cloud computing to be more popular with business organizations and less favored within government agencies. Traditionally, federal government agencies are less risk averse to adopting cloud-computing technology, citing the newness and relative underdevelopment of the cloud services marketplace (Nikolova 2012). However, Daniel (2018) found utilizing cloud technology, state and local government agencies were able to simplify operations while allowing them to improve service to citizens, respond to developing technology needs, and better manage cybersecurity issues.

The purpose of this correlational study was to discover the relationship between the key factors that influence IT directors' decisions to adopt cloud-computing technology within their respective institutions. While seminal research into cloud computing has been published, Gill (2011), Hus and Lin (2016), and Low, Chen, and Wu (2011) found many decision-makers did not understand the factors that contributed to the adoption of cloud computing. Additionally, while there was increased interest among senior decision-makers in adopting cloud computing, concerns over the technology's reliability, stability, and security prevented prompt decision-making (Marston, Li, Bandyopadhyay, Zhang, and Ghalsasi 2011).

This research study used a quantitative approach and attempted to discover the relationship between the key factors that influenced IT directors' or equivalents' decision to adopt cloud-computing technology within their respective institutions. The remainder of this chapter includes a discussion on the chosen research methodology, the research design appropriateness, the research questions and associated hypotheses, the population of the study, a description of the sample size, and a description of how informed consent was obtained. Additionally, there will be a description of the instrumentation used to collect primary data for this study, a discussion on validity and reliability of the study instrument, and finally, a description of the processes used to collect and analyze data.

Research Method and Research Design Appropriateness

Quantitative and qualitative research methods are undertaken by scholars to investigate, explain, and understand the phenomena in such disciplines as psychology, sociology, and public health (Park and Park 2016). To achieve the purpose of this study, a quantitative, nonexperimental, correlational research method was used. Cooper and Schindler (2014) discussed the quantitative methodology as being appropriate when the researcher seeks to measure behavior or attitudes. Additionally, quantitative methods focus on describing, explaining, or predicting a phenomenon by testing or building upon an existing theory as this study uses the TAM as its theoretical foundation (Cooper and Schindler 2014).

In contrast, the qualitative research methodology seeks to provide an in-depth understanding of a particular phenomenon and is traditionally used to inform the researcher on "how" and "why" things happen in the social world (Cooper and Schindler 2014). While qualitative techniques focus on extracting the feelings, emotions, and perceptions of study participants, they are considered fundamentally weak as their results cannot be generalized against a larger population (Cooper and Schindler 2014).

Research designs support the chosen methodology by providing a framework the researcher can use to methodically follow to address the research questions. A research design is the researcher's strategy and structure for the procedures the researcher will follow during the research study, how the researcher collects their data, and the method that will be used to conduct data analysis (Leedy and Ormrod 2014). Several quantitative research designs are available for use within the context of this research study. Babbie (2016) found the experimental design involves taking some action and examining the consequences of those actions, which was grounded in previous empirical research where Campbell and Stanley (1966) posited the experimental research design referred to investigations that involve manipulating the independent variables in the study and observing the resulting effects of that manipulation. In the experimental research design, the researcher intentionally manipulates the independent variable(s) to examine the resulting consequences of establishing a cause-and-effect relationship between the variables. Additionally, as part of the manipulation process, the researcher randomly categorizes and assigns the study participants into experimental and control groups to test the hypothesis (Campbell and Stanley 1966). The degree of control administered during experimentation determines the degree of internal validity when making conclusions about the relationships between the dependent and independent variables (Leedy and Ormrod, 2016). While the goal of this research study was to determine the factors that influence the decision to adopt cloud-computing technologies, the experimental design was not appropriate for this study as according to Campbell and Stanley (1966), the experimental design involves the researcher attempting to control all aspects of the environment and examining the consequences of those actions.

Similar to the experimental design, the quasi-experimental design (sometimes referred to as casual-comparative) involves the use of derived groups to explore the differences between two or more groups or variables (Campbell and Stanley 1966). Shadish, Cook, and Campbell (2001) described the quasi-experimental design as having the same characteristics as the experimental design without the random assignment of groups to conditions. In the experimental design, researchers can control which participants are assigned to the intervention group and which ones are assigned to the comparison group. The strict level of control that is synonymous with the experimental design is lost in the quasi-experimental design. According to Shadish et al. (2001), the cause in a quasi-experiment is manipulated by the researcher before any of the effects are measured. Unfortunately, the quasi-experimental design suffers from several disadvantages. Shadish et al. (2001) postulated the quasi-experimental design is known by researchers to create inconclusive support for false inferences; additionally, as part of the analysis, the researcher has to enumerate alternative explanations to describe any observed effects.

The selected research design for this study was the correlational design, which was found to be most appropriate to effectively measure the statistical relationship among the identified variables that require large data sampling while providing results that can be generalized against large populations (Creswell 2009, Levin 2006). Additionally, the researcher did not attempt to manipulate the independent variables in this study. As such, the correlational design is most appropriate when the researcher intends to explore the relationships among variables that are not manipulated (Fitzgerald, Rumrill, and Schenker 2004). The correlational design was most appropriate for this study to effectively measure the statistical relationship among the identified variables that require a large data sampling while providing results that can be generalized against large populations (Creswell 2009, Levin 2006). According to Fitzgerald, Rumrill, and Schenker (2004), correlation designs are appropriate when researchers want to explore relationships among variables that are not manipulated. Additionally, the casual-comparative design was not appropriate for this study as, according to Schenker and Rumrill (2004), casual-comparative involve the use of derived groups to explore differences between groups or variables.

The ex post facto design was also not appropriate for this study as according to Leedy and Ormrod (2014), the ex post facto designs are more appropriate when a researcher has identified a phenomenon that has already occurred, and the researcher wishes to investigate how an independent variable present before the event may affect the dependent variable.

Research Questions/Hypotheses

RQ1: Is there a correlation between the security effectiveness factor and the decision to adopt cloud computing in a government organization, for-profit business organization, and an academic institution?

H_{01a}: There is no correlation between the security effectiveness factor and the decision to adopt cloud computing in a government organization.

H_{1a}: There is a correlation between the security effectiveness factor and the decision to adopt cloud computing in a government organization.

H_{01b}: There is no correlation between the security effectiveness factor and the decision to adopt cloud computing in a for-profit corporate organization.

H_{1b}: There is a correlation between the security effectiveness factor and the decision to adopt cloud computing in a for-profit corporate organization.

H_{01c}: There is no correlation between the security effectiveness factor and the decision to adopt cloud computing in an academic institution.

H_{1c}: There is a correlation between the security effectiveness factor and the decision to adopt cloud computing in an academic institution.

RQ2: Is there a correlation between the cost-savings factor and the decision to adopt cloud computing in a government organization, for-profit business organization, and an academic institution?

H_{02a}: There is no correlation between the cost-savings factor and the decision to adopt cloud computing in a government organization.

H_{2a}: There is a correlation between the cost-savings factor and the decision to adopt cloud computing in a government organization.

H_{02b}: There is no correlation between the cost-savings factor and the decision to adopt cloud computing in a for-profit corporate organization.

H_{2b}: There is a correlation between the cost-savings factor and the decision to adopt cloud computing in a for-profit corporate organization.

H_{02c}: There is no correlation between the cost-savings factor and the decision to adopt cloud computing in an academic institution.

H_{2c}: There is a correlation between the cost-savings factor and the decision to adopt cloud computing in an academic institution.

RQ3: Is there a correlation between the reliability factor and the decision to adopt cloud computing in a government organization, for-profit business organization, and an academic institution?

H_{03a}: There is no correlation between the reliability factor and the decision to adopt cloud computing in a government organization.

H_{3a}: There is a correlation between the reliability factor and the decision to adopt cloud computing in a government organization.

H_{03b}: There is no correlation between the reliability factor and the decision to adopt cloud computing in a for-profit corporate organization.

H_{3b}: There is a correlation between the reliability factor and the decision to adopt cloud computing in a for-profit corporate organization.

H_{03c}: There is no correlation between the reliability factor and the decision to adopt cloud computing in an academic institution.

H_{3c}: There is a correlation between the reliability factor and the decision to adopt cloud computing in an academic institution.

RQ4: What is the difference between organizations on the decision to adopt cloud computing?

$H_{4:}$ There is a difference between adoption decisions between a government organization, a for-profit corporate organization, and an academic institution.

H_{04}: There is no difference between adoption decisions between a government organization, a for-profit corporate organization, and an academic institution.

Population and Sample

The population targeted for this study was IT directors or equivalent positions within government agencies, for-profit corporate organizations, and academic institutions of higher learning in Central Florida who are either the principal decision-makers or have a significant influence on IT procurement decisions. The survey used for this study determined the factors that contributed to the decision to adopt cloud-computing technologies within each of the identified organizations. Participants for the study consisted of IT directors or subjects who fill an equivalent role (chief information officer, chief technology officer, etc.) working in Central Florida.

The recruitment of suitable respondents for this study utilized a variety of techniques. The researcher leveraged the use of both personal and professional contacts and extended invitations to participate in the study on social media sites such as LinkedIn and Facebook. Additionally, the researcher solicited the services of a survey distribution company, Centiment LLC., to disseminate the survey to participants meeting the criteria for inclusion in the study. In addition to using a purposive sampling method, which, according to Leedy and Ormrod (2016), identified participants meeting the criteria of being a principal decision-maker in IT procurement decisions within their respective organizations, the researcher also employed a snowball sampling technique to assist in obtaining a statistically significant sample of the population.

The sample size for this study was obtained by using the G*Power analysis calculator, using an alpha level = .05, a medium effect size = .30, power = .80 applied against three groups (government agencies, academic institutions, and for-profit corporations) with a one-way ANOVA yielding a total sample population of 111 participants. According to Cohen (1988), the concept of power analysis is used to statistically test if the results of the analysis will yield a significant result. In theory, the power of a statistical test that is applied to research data will lead to the rejection of the null hypothesis, thereby proving the existence of the phenomenon (Cohen 1988). However, the power of a statistical test is dependent upon the test substantiating the existence of the phenomenon, the reliability of the sample size, and its closeness with the relevant population and the effect size (Cohen 1988). While the study was limited to Central Florida, it would be beneficial to collect data from the entire targeted population; however, it would be impractical and unnecessary to use too large a population (Christen et al. 2016). See Appendix A for the G* Power computation.

Informed Consent and Confidentiality

Obtaining informed consent from study participants constitutes an integral part of the practice of academic research (Sivanadarajah, El-Daly, Mamarelis, Sohail, and Bates 2017). Once approval was granted from the university institutional review board, all study participants were informed about the nature and purpose of the research. Participants were presented with a detailed informed consent form on the front page of the online survey. Study participants were notified their participation in the study was voluntary, and they may decline to continue participation in the study at any point should they desire. Study participants were notified their anonymity would be assured as their names will not be used in any portion of the study. The survey procedures were highlighted along with any expected benefits that may result from the study before the participant began taking the survey.

According to Leedy and Ormrod (2016), confidentiality in research is an ethical issue the researcher must remain cognizant about as information is collected, stored, and utilized in the performance of the study. The surveys used in this study to collect data, at a minimum, captured the respondents' demographic data to include name and age range. However, the survey was not designed or set up to collect personally identifiable information such as addresses, phone numbers, birth dates, or Social Security numbers. To ensure the safekeeping of the participants' identity, responses and coding research artifacts were kept in an encrypted folder on an external storage device. The researcher is the only person with an encrypted password to access the data. Although demographic data was captured and associated with the responses, participant names were not linked back to responses to maintain anonymity. Upon completion of the study and acceptance of the results by the university, all associated electronic files will continue to be maintained for a period of three years per university research retention polices. Once the research files have reached the end of life per university polices for retention, all files associated with this study will be destroyed following university policy for the destruction of research materials.

Instrumentation

The survey instrument that was used for this study to collect data was the self-administered questionnaire adapted from Ali et al. (2016). According to Ali et al. (2016), "the survey was developed based on previous literature on technological and organizations studies and the findings from a qualitative study" (p. 313). The survey instrument utilized a 5-point Likert-type scale consisting of close-ended questions targeted toward IT decision-makers in government agencies, for-profit corporate organizations, and academic institutions. Participants answered questions on the survey presented in ordinal data form with responses ranging from strongly disagree (1) to strongly agree (5). This assertion is supported by Grün and Dolnicar (2016), who posited that survey data is typically rendered as ordinal in nature and is subject to response styles, which can be problematic in survey research. While the survey responses were input into the online survey tool SurveyMonkey, the data was not aggregated using analysis tools within SurveyMonkey. The raw data was retrieved and input into SPSS for statistical analysis. The survey results were converted from their ordinal scale points to the numeric equivalent and analyzed for normalcy to determine if there was a central tendency. Upon completion of analysis, the final variable was designated as a nominal variable. The following table depicts how the instrumentation items aligned to the research questions (table 1).

Table 1. ***Research Question to Survey Questions Linking Table***

RESEARCH QUESTION	LINKED SURVEY QUESTIONS
RQ1: Is there a correlation between the security effectiveness factor and the decision to adopt cloud computing in a government agency, for-profit business organization, and an academic institution?	**Likert-type scale using categories of (strongly disagree, disagree, neither agree or disagree, agree, and strongly agree)** - Data security is a major concern for organizations considering to adopt cloud computing. - Cloud computing provides a sufficient security transfer channel during the process of mass data interchange. - Loss of control over data and applications influences the adoption decision. - Cloud computing provides a secure service. - Security concerns are not an issue with cloud computing.

RQ2: Is there a correlation between the cost-savings factor and the decision to adopt cloud computing in a government agency, for-profit business organization, and an academic institution?	**Likert-type scale using categories of (strongly disagree, disagree, neither agree or disagree, agree, and strongly agree)** - Cloud computing decreases the investment cost in new IT infrastructure. - Cloud computing is more cost-effective compared with the other IS technologies. - Cloud computing reduces the costs of system upgrades. - Cloud computing may increase the ability of an organization to adapt rapidly and cost efficiently in response to changes in the government, academic, or business environment. - Cloud computing reduces the total cost of operational processes.
RQ3: Is there a correlation between the reliability factor and the decision to adopt cloud computing in a government agency, for-profit business organization, and an academic institution?	**Likert-type scale using categories of (strongly disagree, disagree, neither agree or disagree, agree, and strongly agree)** - Cloud computing provides a reliable service with high availability. - Cloud computing can/may enable better communications between our suppliers and customers. - Cloud computing can create a flexible environment to operate in. - Cloud computing can provide access to cloud services from various client devices (e.g., thick clients, thin clients, laptops, tablets, smartphones, etc.) in our environment. - Cloud computing can enable the organization to respond quickly to customer requests.

Validity and Reliability

The survey used in this study was adopted from Ali et al. (2016), who developed the survey using a two-stage process to ensure accuracy. According to Ali et al. (2016), the first stage involved a pilot or pretest carried out at one site consisting of thirty IT members to test the survey's validity, identify any issues with the questions, and make improvements on the survey. As a result of the pilot testing, the initial survey questions were adjusted to improve the design of the questionnaire and eliminate questions that were not beneficial to the study (Ali et al. 2016).

According to Neuman (2006), reliability in research is described as having the characteristics of dependability and consistency throughout the entire research process. In the study, the survey instrument was adopted from the researcher who conducted the pilot study to evaluate the survey's validity, reliability, and predictability (Ali et al. 2016). The pilot study only sampled thirty IT professionals on the chosen study site with a 70 percent response rate. The final questionnaire was updated based on feedback received from the pilot study, which revealed Cronbach alpha values greater than .80 for each of the survey constructs (Ali et al. 2016).

To ensure reliability in this study, the researcher chose to use the Cronbach's coefficient of reliability on each of the constructs used in this study. Cronbach's alpha is typically used when determining reliability as it relates to internal consistency when using combined measurements (Inal, Yilmaz Koğar, Demirdüzen, and Gelbal 2017). Additionally, Cronbach's alpha is a test reliability index calculated using classical test theory approaches (Zumbo 1999). Cronbach's alpha measures how closely related a set of variables are in a group. Typically, a Cronbach alpha of 0.70 or higher is considered acceptable (Dinjens, Senden, Heyligers, and Grimm 2014).

Previous studies using Likert scaled surveys have also used the Cronbach alpha to ensure reliability. Shroff, Ting, and Lam (2019) studied the development of the Technology-Enabled Active Learning Inventory (TEAL) survey instrument used to measure student perceptions of technology-enabled active learning. During their study, the TEAL survey demonstrated internal consistency and reliability validated by Cronbach's alpha scores ranging from 0.83 to 0.88, indicating a high degree of reliability and internal consistency. Further validating Cronbach's alpha's use in empirical research, Huston (2018) used Cronbach's alpha to demonstrate reliability when studying the factors associated with entrepreneurial intentions in pharmacy students. Using a thirty-one–item Likert scale survey, Huston (2018) found the chosen instrument yielded a Cronbach alpha = 0.89, exhibiting a significant degree of reliability. Finally, before conducting their survey to analyze the factors affecting employee satisfaction in Pakistan, Rukh, Choudhary, and Abbasi (2015) tested the reliability of their Likert scaled questionnaire using Cronbach's alpha. Using a fifty-seven–item questionnaire, test results revealed an overall Cronbach's alpha reliability coefficient of 0.966 (Rukh et al. 2015).

Data Collection

The process of data collection in research involves the researcher making decisions as it relates to the type of research being done, sampling methods being employed, and the type of variables used in the study. However, before data analysis can occur, data must first be collected as collecting data is the foundation of all empirical research (Aguinis, Hill, and Bailey 2019). Data for this study was collected using several collection methods to include online distribution of surveys via SurveyMonkey, the use of the survey distribution service Sentiment LLC., and distributing the survey questionnaire on social media sites like LinkedIn and Facebook to participants meeting the required criteria for inclusion in the survey. The researcher provided potential participants the link for the survey, which contained the research consent form and instructions for completing the survey. To provide survey participants with a certain degree of anonymity, the survey instrument did not collect personally identifiable information to ensure responses did not correlate back to a specific respondent. Although some survey data initially resided on SurveyMonkey servers for initial data analysis, the aggregated raw data was downloaded into an Excel spreadsheet on the researchers' computer and deleted from the SurveyMonkey website. All data was input or uploaded into the SPSS software for processing and statistical analysis. The completed survey responses were used for later correlational analysis.

To ensure the safekeeping of participant responses, all survey data was stored in a protected password-enabled folder on the researcher's computer. The researcher was the only person with the encrypted password to access the data. Upon completion of the study and acceptance of the results by the university, all associated electronic files will continue to be maintained for a period of three years in accordance with university research retention polices. Once the research files have reached the end of life per university polices for retention, all files associated with this study will be destroyed in accordance with university policy for the destruction of research materials.

Data Analysis

The purpose of this quantitative correlational study was to determine the factors that influenced the decision to adopt cloud-computing technologies between government agencies, for-profit corporate organizations, and academic institutions. Since this study used multiple variables, a technique involving the use of multivariate statistics was used to analyze the data. The term *multivariate statistics* traditionally includes all statistics where there are more than two variables simultaneously analyzed. In situations where the problems involve three or more variables, they are inherently multidimensional and require the use of multivariate data analysis (such as correlation, ANOVA, and multiple regression).

The specific multivariate techniques used in this study were the one-way ANOVA to compare the means among the three groups and the Pearson moment correlation, more commonly referred to as the Pearson r correlation, to measure how closely related the variables are within each group. The one-way ANOVA is used to compare the means of multiple groups that result in the outcome of variations being examined and divided into two parts, the difference between the groups being compared and the residual outcomes that cannot be explained (Lewis-Beck, Bryman, and Futing Liao 2004). According to Wall (2015), the Pearson r correlation is the most commonly used statistic to measure how closely two variables are related. The equation used to calculate this statistic relates two sets of scores for two different measures. The resulting value is a single number referred to as the correlation coefficient designated by the letter *r* (Wall 2015). According to Tokunaga (2016), the Pearson correlation coefficient measures the linear relationship between two continuous variables. The *r* statistic is designed to measure nature, direction, and strength. The *r* statistic assesses the degree of the linear relationship between two variables as long as the relationship between the variables is linear (Tokunaga 2016). The direction of the relationship is illustrated by a (+/-) as either being a positive or negative Pearson correlation coefficient; positive values represent a positive relationship, and negative values represent a negative relationship. The strength of the relationship is depicted by a numeric value for *r* ranging from -1.00 to +1.00. According to Tokunaga (2016), the strongest and perfect relationship is represented by a Pearson correlation coefficient of either -1.00 or +1.00.

Data was input into the IBM SPSS software application where inferential and descriptive statistics were run to examine the variables, aid in understanding, and to summarize the numerical data. Additionally, applying the Pearson r coefficient correlational function within SPSS allowed for the interpretation of the data as well as a graphical representation via a scatterplot.

Summary

This correlational study focused on surveying IT directors or persons in equivalent-level positions within a government agency, for-profit corporate organization, and an academic institution to determine the relationship between the factors that influenced the decision to adopt cloud-computing technologies. The discussion in this chapter focused on the suitability of the quantitative methodology as the best approach to determine if there is a statistical significance between the identified variables used within the study. Additionally, there was a discussion on the relevance of the selected correlational research design as the most appropriate for this study as correlational designs effectively measure statistical relationships while enabling the results to be generalized against large populations (Creswell 2009, Levin 2006).

This chapter further explored how the study population and sample were derived while describing the chosen survey instrument used to collect the data. Empirical evidence was presented, legitimizing the validity and reliability of the survey instrument. Finally, the chapter concluded with an examination of the data analysis technique used to decipher the data.

Chapter 4 will provide the research findings and an in-depth interpretation of the analysis.

4

Analysis and Results

The purpose of this correlational study was to discover the relationship between the key factors (security effectiveness, cost savings, and reliability) that influence IT directors' decisions to adopt cloud-computing technology within their respective institutions. According to Ghazizadeh, Lee, and Boyle (2012), previous research revealed the Technology Acceptance Model (TAM) has been used in empirical research and validated across various technologies and populations. Furthermore, the TAM has served as a guide for researchers in identifying the internal and external factors impacting the decision on whether to adopt cloud computing. The results of this study will contribute to the body of knowledge in cloud computing by providing salient information for decision-makers regarding viable approaches they can employ to increase acceptance of cloud computing within their respective organizations.

The survey instrument that was used for this study to collect data was the self-administered questionnaire adapted from Ali et al. (2016). According to Ali et al. (2016), "the survey was developed based on previous literature on technological and organizations studies and the findings from a qualitative study" (p. 313). The survey instrument utilized a 5-point Likert-type scale consisting of close-ended questions targeted toward IT decision-makers in government agencies, for-profit corporate organizations, and academic institutions. Participants answered questions on the survey presented in ordinal data form with responses ranging from strongly disagree (1) to strongly agree (5).

Research Questions/Hypotheses

The specific research questions and hypotheses that guided this study to discover the relationship between the key factors that IT decision-makers focus on when they make their decision to adopt cloud-computing technology were the following:

RQ1: Is there a correlation between the security effectiveness factor and the decision to adopt cloud computing in a government organization, for-profit business organization, and an academic institution?

H_{01a}: There is no correlation between the security effectiveness factor and the decision to adopt cloud computing in a government organization.

H_{1a}: There is a correlation between the security effectiveness factor and the decision to adopt cloud computing in a government organization.

H_{01b}: There is no correlation between the security effectiveness factor and the decision to adopt cloud computing in a for-profit corporate organization.

H_{1b}: There is a correlation between the security effectiveness factor and the decision to adopt cloud computing in a for-profit corporate organization.

H_{01c}: There is no correlation between the security effectiveness factor and the decision to adopt cloud computing in an academic institution.

H_{1c}: There is a correlation between the security effectiveness factor and the decision to adopt cloud computing in an academic institution.

RQ2: Is there a correlation between the cost-savings factor and the decision to adopt cloud computing in a government organization, for-profit business organization, and an academic institution?

H_{02a}: There is no correlation between the cost-savings factor and the decision to adopt cloud computing in a government organization.

H_{2a}: There is a correlation between the cost-savings factor and the decision to adopt cloud computing in a government organization.

H_{02b}: There is no correlation between the cost-savings factor and the decision to adopt cloud computing in a for-profit corporate organization.

H_{2b}: There is a correlation between the cost-savings factor and the decision to adopt cloud computing in a for-profit corporate organization.

H_{02c}: There is no correlation between the cost-savings factor and the decision to adopt cloud computing in an academic institution.

H_{2c}: There is a correlation between the cost-savings factor and the decision to adopt cloud computing in an academic institution.

RQ3: Is there a correlation between the reliability factor and the decision to adopt cloud computing in a government organization, for-profit business organization, and an academic institution?

H_{03a}: There is no correlation between the reliability factor and the decision to adopt cloud computing in a government organization.

H_{3a}: There is a correlation between the reliability factor and the decision to adopt cloud computing in a government organization.

H_{03b}: There is no correlation between the reliability factor and the decision to adopt cloud computing in a for-profit corporate organization.

H_{3b}: There is a correlation between the reliability factor and the decision to adopt cloud computing in a for-profit corporate organization.

H_{03c}: There is no correlation between the reliability factor and the decision to adopt cloud computing in an academic institution.

H_{3c}: There is a correlation between the reliability factor and the decision to adopt cloud computing in an academic institution.

RQ4: What is the difference between organizations on the decision to adopt cloud computing?

H_4: There is a difference between adoption decisions between a government organization, a for-profit corporate organization, and an academic institution.

H_{04}: There is no difference between adoption decisions between a government organization, a for-profit corporate organization, and an academic institution.

Data Collection

The survey questionnaire was transferred to SurveyMonkey to begin the data collection process. The researcher provided potential participants the link for the survey, which contained the research consent form and instructions for completing the survey. The survey questionnaire was distributed by the survey distribution service Centiment LLC. and posted by the researcher on social media sites like LinkedIn and Facebook to participants meeting the required criteria for inclusion in the survey. To provide survey participants with anonymity, the survey instrument did not collect personally identifiable information to ensure responses did not correlate back to a specific respondent. Although initial survey data resided on SurveyMonkey servers, the aggregated raw data was exported into an Excel spreadsheet on the researcher's computer and deleted from the SurveyMonkey website. All data was imported into SPSS where statistical analysis was completed.

The target population for this study were IT decision-makers. responsible for procurement actions within government agencies, for-profit corporate organizations, and academic institutions of higher learning in Central Florida. Chapter 4 is organized by a discussion of the data preparation, sample demographics, reliability analysis, descriptive statistics, research question/hypothesis testing, and a summary of the results. Data was analyzed with SPSS 23 for Windows. The following provides a discussion of the data analysis.

Demographics

The sample consisted of 111 participants distributed among government (31.5%, n = 35), academic (34.2%, n = 38), and corporate (for-profit) (34.2%, n = 38) organizations. Regarding educational level, 16.2% (*n* = 18) were undergraduates, 36.9% (*n* = 41) were graduates, 44.1% (*n* = 49) were postgraduates, and 2.7% (*n* = 3) were "other." Participants described their roles as chief information officers (CIO) (9.9%, *n* = 11), IT directors (47.7%, *n* = 53), and chief technology officers (CTO) (4.5%, *n* = 5) to name a few. Roles are presented in table 2.

Table 2. Role

Variable	*n*	%
CIO	11	9.9
IT Director	53	47.7
Equivalent level position	17	15.3
DEP CIO	7	6.3
CTO	5	4.5
Other	18	16.2
Total	111	100.0

Concerning age, the largest group (47.7%, *n* = 53) was comprised of participants 35–44 years of age. The next largest age group (23.4%, *n* = 26) was 25–34. However, the third-largest group of respondents (6.3%, *n* = 7) were individuals 18–24 years of age. Age is presented in table 3.

Table 3. Age

Age	*n*	%	Cumulative %
18–24	7	6.3	6.3
25–34	26	23.4	29.7
35–44	53	47.7	77.5
45–54	18	16.2	93.7
55–64	6	5.4	99.1
65–74	1	0.9	100.0
Total	111	100.0	

Regarding gender, males (59.5%, n = 66) were in the majority and females (40.5%, n = 45) were in the minority. Participants were asked if their organizations used cloud-computing technology. Most (95.5%, n = 106) selected yes, whereas 4.5% (n = 5) selected no. Participants were then asked what cloud-based services/applications were used by their organizations. Respondents were asked to check all that applied. For example, 66.7% of participants (n = 74) used software as a service, 44.1% (n = 49) used platform as a service, and 35.1% (n = 39) used "other platform." Cloud-based services/applications used in participant organizations are presented in table 4.

Table 4. Cloud-Based Services/Applications Use in Participant Organizations

Cloud-Based Services/ Applications	No		Yes	
	n	%	*n*	%
Software as a Service	37	33.3%	74	66.7%
Platform as a Service	62	55.9%	49	44.1%
Infrastructure as a Service	72	64.9%	39	35.1%
Other Platform	107	96.4%	4	3.6%
Email	33	29.7%	78	70.3%
SMS/Text Messaging	58	52.3%	53	47.7%
Telephone Services/VoIP	61	55.0%	50	45.0%
Office Applications	59	53.2%	52	46.8%
Web Conferencing	48	43.2%	63	56.8%
Customer Relationship Management (CRM)	74	66.7%	37	33.3%
Virtual Computers/Servers	65	58.6%	46	41.4%
Data Backup/Storage	53	47.7%	58	52.3%
Disaster Recovery	93	83.8%	18	16.2%
Website Hosting	75	67.6%	36	32.4%
Operating System	71	64.0%	40	36.0%
E-commerce Cloud-Based Services	92	82.9%	19	17.1%
Remote Access/VPN	86	77.5%	25	22.5%

Social Networking	77	69.4%	34	30.6%
Payroll	80	72.1%	31	27.9%
Billing and Invoicing	87	78.4%	24	21.6%

A bar graph was created to depict the most frequently used cloud-based applications by summing the total values (frequencies) for each cloud-based service/application. Then the frequencies were sorted in descending order, which facilitated ease of interpretation for most frequent use to least frequent use of each service/application. For instance, the most frequent cloud-based services/applications used in participant organizations were email (70.3%, n = 73), software as a service (66.7%, n = 74), and web conferencing (56.8%, n = 63). The least frequent cloud-based services/applications were "other platforms" (3.6%, n = 4), disaster recovery (16.2%, n = 18), and e-commerce cloud-based services (17.1%, n = 19) (see figure 6).

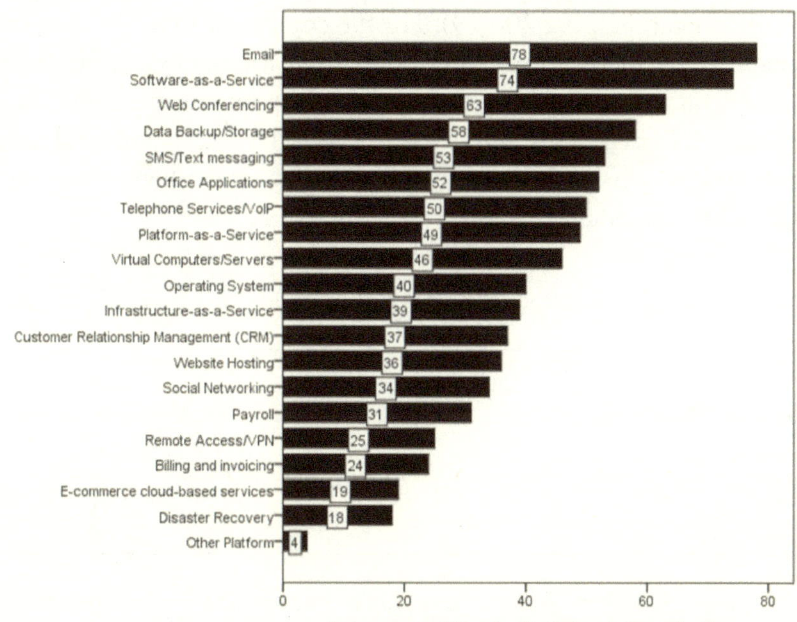

Figure 6. Cloud-Based Services/Applications Used by Participant Organizations

Data Analysis

Data Preparation

The data was exported from SurveyMonkey to SPSS for analysis. The data was examined visually for missing data. There was no missing data. Columns and variables were relabeled for ease of reading. Reliability analyses was conducted, and composite scores for the variables of interest were computed.

Instrument Reliability for Sample

Instrument reliability for the sample was tested with Cronbach's alpha. For the security factor, the initial reliability was poor (α = .59). An inter-item analysis was conducted on the data. It was determined that the reliability for the security factor could be improved by 10 points if one item (*security concerns are not an issue with cloud computing*) was deleted from the subscale. Therefore, the reliability for the security factor changed from poor to questionable (α = .69), bordering on acceptable (.70–.79). For the cost factor (α = .75) and the reliability factor (α = .76), the internal consistency was acceptable. The overall internal consistency for the instrument was good (α = .88) for all fifteen items, but it was .89 with the aforementioned item excluded, which is still good but bordering on excellent (.90–.99). Reliability coefficients are presented in table 5.

Table 5. Reliability Coefficients

Variable	*N* of Items	Cronbach's alpha	Interpretation
Security Factor*	4	.689	Questionable
Cost Factor	5	.752	Acceptable
Reliability Factor	5	.758	Acceptable
Decision to Adopt*	14	.888	Good

**Note*. One item was dropped due to poor reliability. Interpretations are based on generally accepted criteria (DeVellis 2012).

The survey instrument utilized a 5-point Likert-type scale consisting of 1(*strongly disagree*) to 5 (*strongly agree*). Scores for the variables of interest were computed by calculating the mean responses for each variable. Therefore, scores could range from 1 to 5. In actuality, for the security factor, scores ranged from 2.00 to 5.00 (M = 4.20, SD = 0.59). For each variable of interest, the mean value can be rounded to 4.00, which corresponds to "agree" on the survey. This suggests general agreement that IT decision-makers are influenced by the security factor, the cost factor, and the reliability factors when they decide to adopt cloud computing. Descriptive statistics are presented in table 6.

Table 6. Descriptive Statistics

Variable	*n*	*Minimum*	*Maximum*	*M*	*SD*
Security Factor	111	2.00	5.00	4.20	0.59
Cost Factor	111	1.00	5.00	4.05	0.63
Reliability Factor	111	1.00	5.00	4.15	0.53
Decision to Adopt	111	1.29	4.93	4.13	0.52

The data was screened for normality with skewness and kurtosis statistics, the Shapiro-Wilk Test of Normality, and illustrated with histograms. In SPSS, distributions are normal if the absolute values of their skewness and kurtosis statistics are less than two times their standard errors (George and Mallery 2010). Based on these criteria, all the distributions were not normal. Skewness and kurtosis coefficients are presented in table 7.

Table 7. Skewness and Kurtosis Coefficients

	Skewness		Kurtosis	
Variable	Statistic	Std. Error	Statistic	Std. Error
Security Factor	-.816	.229	.908	.455
Cost Factor	-.910	.229	3.68	.455
Reliability Factor	-1.80	.229	9.66	.455
Decision to Adopt	-1.37	.229	6.60	.455

The Shapiro-Wilk Test of Normality compares the sample distribution to a theoretical normal distribution. If there is a significant difference ($p < .05$), then the distribution is not normal. The Shapiro-Wilk Test of Normality indicated that all the distributions for the variables of interest were not normal. Shapiro-Wilk Test of Normality results are presented in table 8.

Table 8. Shapiro-Wilk Test of Normality

Variable	Shapiro-Wilk Statistic	*df*	*p*
Security Factor	.931	111	.000
Cost Factor	.921	111	.000
Reliability Factor	.853	111	.000
Decision to Adopt	.904	111	.000

For the security factor, the skewness coefficient was 3.56 times the standard error. The kurtosis was 2.00 times the standard error. The Shapiro-Wilk Test of Normality also indicated that the distribution for the security factor was not normal ($p < .05$). The histogram for the security factor is presented in figure 7.

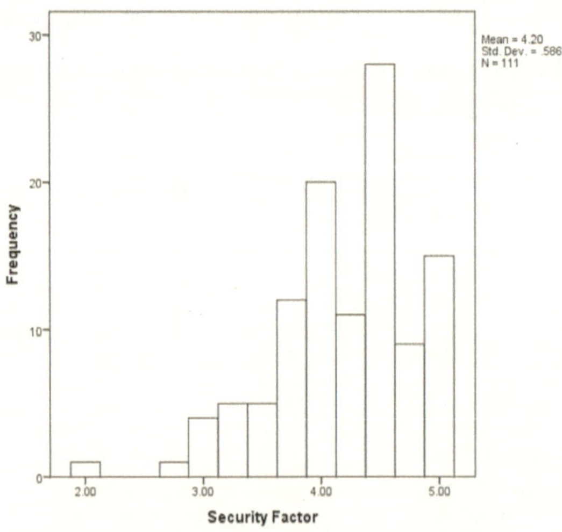

Figure 7. Histogram of the Security Factor

The data were also screened for statistical outliers with stem and leaf plots and also with box and whisker plots. Outliers are indicated as points outside the whiskers in box and whisker plots. They are determined mathematically when they fall outside of 1.5 times the interquartile range (IQR). For the security factor, the median = 4.25 and the IQR = 0.75. There were two statistical outliers ≤ 2.8. The box and whisker plot for the security factor is presented in figure 8.

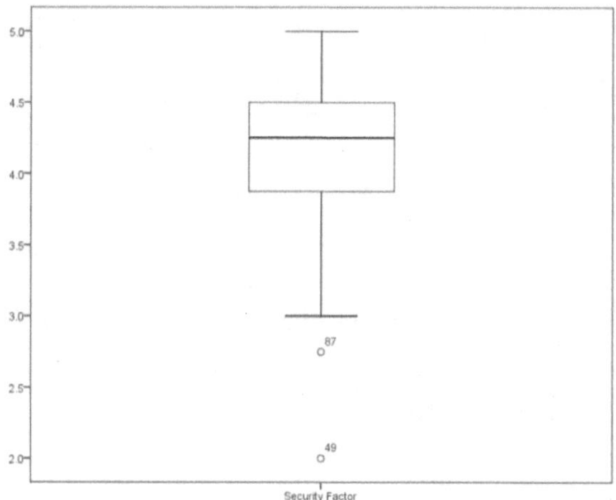

Figure 8. Box and Whisker Plot for the Security Factor

For the cost factor, the skewness coefficient was 3.97 times the standard error. The kurtosis was 8.09 times the standard error. The Shapiro-Wilk Test of Normality also indicated that the distribution for the cost factor was not normal ($p < .05$). The histogram for the cost factor is presented in figure 9.

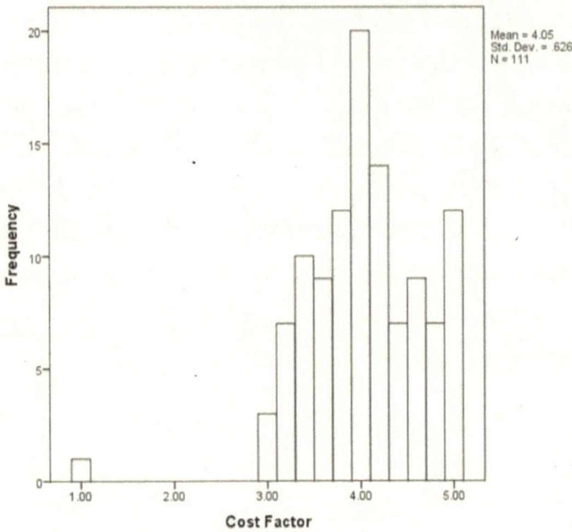

Figure 9. Histogram of Cost Factor

For the cost factor, the median = 4.00 and the IQR = 1.00. There was one statistical outlier ≤ 1.0. The box and whisker plot for the cost factor is presented in figure 10.

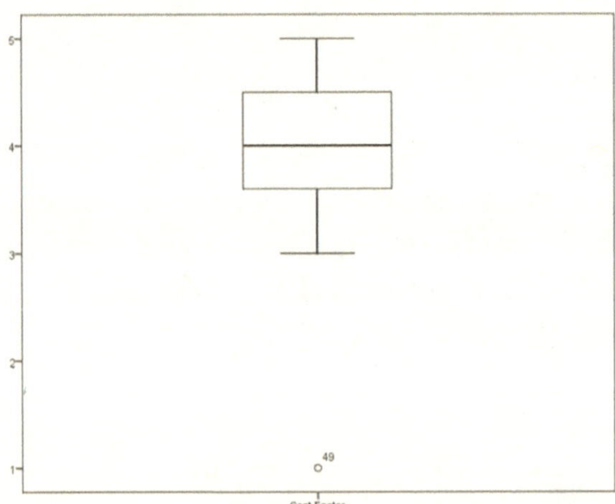

Figure 10. Box and Whisker Plot for Cost Factor

For the reliability factor, the skewness coefficient was 7.86 times the standard error. The kurtosis was 21.23 times the standard error. The Shapiro-Wilk Test of Normality also indicated that the distribution for the reliability factor was not normal ($p < .05$). The histogram for the reliability factor is presented in figure 11.

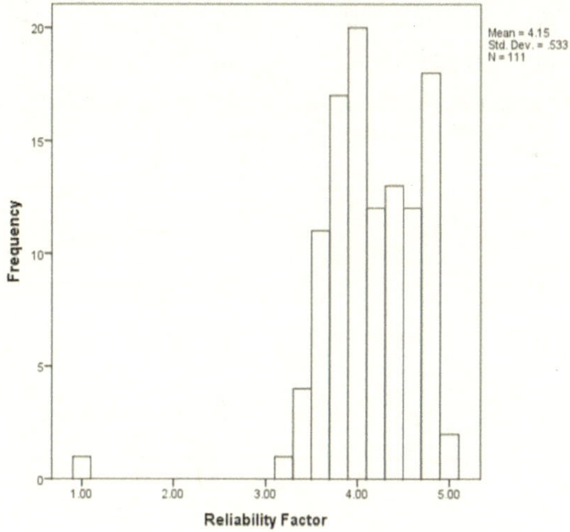

Figure 11. Histogram of Reliability Factor

For the reliability factor, the median = 4.20 and the IQR = 0.80. There was one statistical outlier ≤ 1.0. The box and whisker plot for the reliability factor is presented in figure 12.

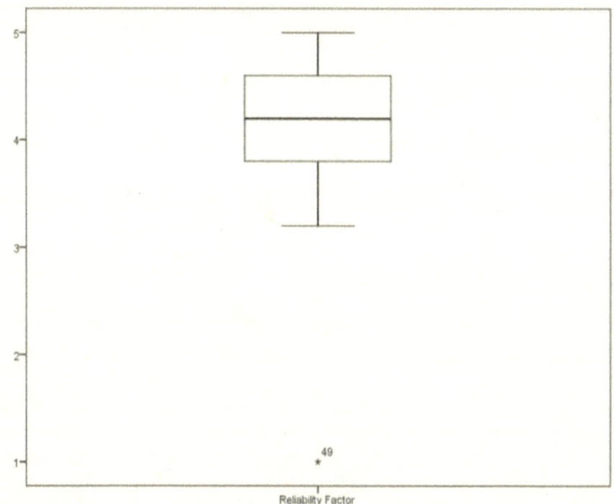

Figure 12. Box and Whisker Plot for Reliability Factor

For the decision to adopt cloud-computing technology, the skewness coefficient was 5.98 times the standard error. The kurtosis was 14.51 times the standard error. The Shapiro-Wilk Test of Normality also indicated that the distribution for the decision to adopt cloud computing technology was not normal ($p < .05$). The histogram for the decision to adopt cloud computing technology is presented in figure 13.

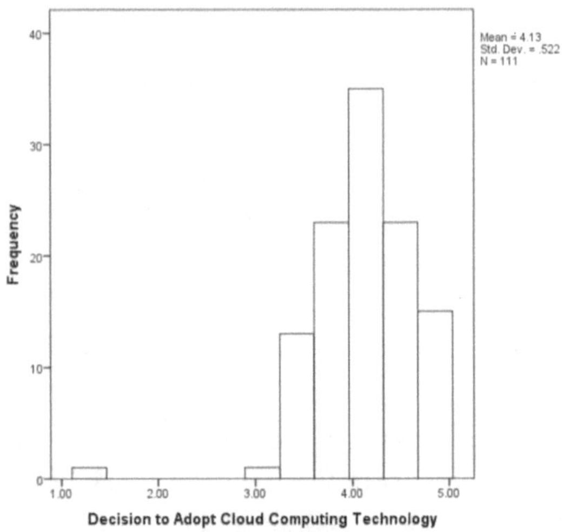

Figure 13. Histogram of Decision to Adopt Cloud-Computing Technology

For the decision to adopt cloud-computing technology, the median = 4.14 and the IQR = 0.71. There was one statistical outlier ≤ 1.29. The box and whisker plot for the reliability factor is presented in figure 14.

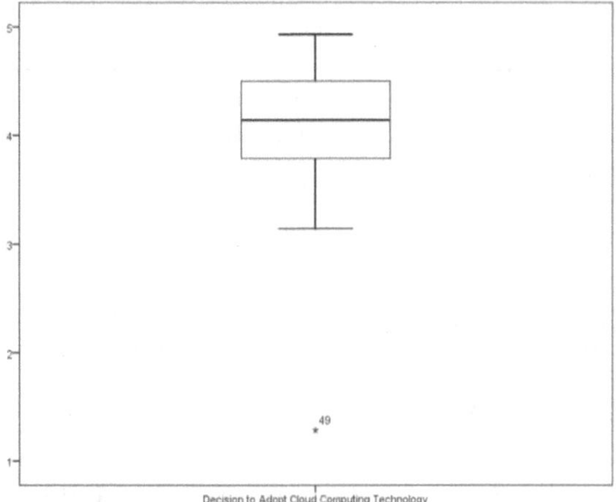

Figure 14. Box and Whisker Plot for Decision to Adopt Cloud-Computing Technology

To summarize the results of the data screening, the distributions were not normal based on the skewness and kurtosis coefficients and also the Shapiro-Wilk Test of Normality. There were also outliers present in the distributions. To address the statistical outliers, z-scores were computed. Any z-score that was greater than 3.00 was considered an outlier and was a candidate for exclusion from further analyses. One case met this criteria and was therefore excluded from further analyses. After the outlier was removed, the data were more normalized. Using the explore feature in SPSS, skewness and kurtosis statistics were generated again. Skewness values were less than two times their standard errors for all but one variable, security factor. For security factor, the skewness was 2.23 times its standard error. Although it was above the threshold for normality, it was only slightly above the threshold and was a reduction from the 3.56 factor initially obtained. Histograms for the data after the outlier was removed are presented in appendix E. Skewness and kurtosis statistics after the outlier was removed are provided in table 9.

Table 9. Skewness and Kurtosis Statistics after Outlier Removal

	Skewness		Kurtosis	
Variable	Statistic	Std. Error	Statistic	Std. Error
Security Factor	-.512	.230	-.284	.457
Cost Factor	.068	.230	-.848	.457
Reliability Factor	.017	.230	-1.04	.457
Decision to Adopt	.018	.230	-.770	.457

Research Questions and Hypothesis Testing

Four research questions were investigated. To answer the first three research questions, the data set was split by the three organizational types—government, academic, and corporate (for-profit)—using the split file command. The Pearson r was conducted on the data. Splitting the file before conducting the analyses made it possible to generate correlation coefficients for the first three research questions simultaneously. A correlation matrix for all the variables of interest for each organization type is presented in appendix F. Based on the values obtained from the correlational analyses in appendix F, smaller tables were created for each research question to facilitate ease of interpretation.

Is there a correlation between the security effectiveness factor and the decision to adopt cloud computing in a government organization, for-profit business organization, and an academic institution? Correlation coefficients and significance levels for research question 1 are presented in table 10.

Table 10. Security Effectiveness and the Decision to Adopt Cloud Computing

	Government N=35	Academic N = 37	Corporate (For-Profit) N = 38
r	.834	.863	.744
p	< .001	< .001	< .001

**Note.* Predictor variable = Security effectiveness, criterion variable = Decision to adopt cloud computing technology, two-tailed.

There was a significant, strong, positive relationship between the security effectiveness factor and the decision to adopt cloud computing in a government organization, $r(33) = .83$, $p < .001$, two-tailed. The coefficient of determination (r^2) = .6889, which means that 68.9% of the variance in cloud-computing technology can be explained by the security effectiveness factor for government organizations.

There was a significant, strong, positive relationship between the security effectiveness factor and the decision to adopt cloud computing in an academic organization, $r(35) = .86$, $p < .001$, two-tailed. The coefficient of determination (r^2) = .7396, which means that 74% of the variance in cloud-computing technology can be explained by the security effectiveness factor in academic organizations.

There was a significant, strong, positive relationship between the security effectiveness factor and the decision to adopt cloud computing in a corporate (for-profit) organization, $r(36) = .74$, $p < .001$, two-tailed. The coefficient of determination (r^2) = .5476, which means that 54.8% of the variance in cloud-computing technology can be explained by the security effectiveness factor in corporate (for-profit) organizations. Scatterplots of these relationships are presented in appendix G.

$\mathbf{H_{01a}}$ stated that there is no correlation between the security effectiveness factor and the decision to adopt cloud computing in a government organization. There was a significant, strong, positive relationship between the security effectiveness factor and the decision to adopt cloud computing in a government organization, $r(33) = .83$, $p < .001$, two-tailed. Therefore, the null hypothesis was rejected.

$\mathbf{H_{01b}}$ stated that there is no correlation between the security effectiveness factor and the decision to adopt cloud computing in a for-profit corporate organization. There was a significant, strong, positive relationship between the security effectiveness factor and the decision to adopt cloud computing in a corporate (for-profit) organization, $r(36) = .74$, $p < .001$, two-tailed. Therefore, the null hypothesis was rejected.

$\mathbf{H_{01c}}$ stated that there is no correlation between the security effectiveness factor and the decision to adopt cloud computing in an academic institution. There was a significant, strong, positive relationship between the security effectiveness factor and the decision to adopt cloud computing in an academic organization, $r(35) = .86, p < .001$, two-tailed. Therefore, the null hypothesis was rejected.

Is there a correlation between the cost-savings factor and the decision to adopt cloud computing in a government organization, for-profit business organization, and an academic institution? Correlation coefficients and significance levels for research question 2 are presented in table 11.

Table 11. Cost-Savings Factor and the Decision to Adopt Cloud Computing

	Government N=35	Academic N = 37	Corporate (For-Profit) N = 38
r	.903	.889	.932
p	< .001	< .001	< .001

**Note.* Predictor variable = Cost-savings factor, criterion variable = Decision to adopt cloud computing technology, two-tailed.

There was a significant, strong, positive relationship between the cost-savings factor and the decision to adopt cloud computing in a government organization, $r(33) = .90, p < .001$, two-tailed. The coefficient of determination (r^2) = .81, which means that 81% of the variance in cloud-computing technology can be explained by the cost-savings factor for government organizations.

There was a significant, strong, positive relationship between the cost-savings factor and the decision to adopt cloud computing in an academic organization, $r(35) = .89, p < .001$, two-tailed. The coefficient of determination (r^2) = .7921, which means that 79.2% of the variance in cloud-computing technology can be explained by the cost-savings factor in academic organizations.

There was a significant, strong, positive relationship between the cost-savings factor and the decision to adopt cloud computing in a corporate (for-profit) organization, $r(36) = .93, p < .001$, two-tailed. The coefficient of determination (r^2) = .8649, which means that 86.5% of the variance in cloud-computing technology can be explained by the cost-savings factor in corporate (for-profit) organizations. Scatterplots of these relationships are presented in appendix H.

$\mathbf{H_{02a}}$ stated that there is no correlation between the cost-savings factor and the decision to adopt cloud computing in a government organization. There was a significant, strong, positive relationship between the cost-savings factor and the decision to adopt cloud computing in a government organization, $r(33) = .90$, $p < .001$, two-tailed. Therefore, the null hypothesis was rejected.

$\mathbf{H_{02b}}$ stated that there is no correlation between the cost-savings factor and the decision to adopt cloud computing in a for-profit corporate organization. There was a significant, strong, positive relationship between the cost-savings factor and the decision to adopt cloud computing in a corporate (for-profit) organization, $r(36) = .93$, $p < .001$, two-tailed. Therefore, the null hypothesis was rejected.

$\mathbf{H_{02c}}$ stated that there is no correlation between the cost-savings factor and the decision to adopt cloud computing in an academic institution. There was a significant, strong, positive relationship between the cost-savings factor and the decision to adopt cloud computing in an academic organization, $r(35) = .89$, $p < .001$, two-tailed. Therefore, the null hypothesis was rejected.

Is there a correlation between the reliability factor and the decision to adopt cloud computing in a government organization, for-profit business organization, and an academic institution? Correlation coefficients and significance levels for research question 3 are presented in table 12.

Table 12. Reliability Factor and the Decision to Adopt Cloud Computing

	Government *N* =35	Academic *N* = 37	Corporate (For-Profit) *N* = 38
r	.884	.856	.873
p	< .001	< .001	< .001

**Note.* Predictor variable = Reliability factor, criterion variable = Decision to adopt cloud computing technology, two-tailed.

There was a significant, strong, positive relationship between the reliability factor and the decision to adopt cloud computing in a government organization, $r(33) = .88$, $p < .001$, two-tailed. The coefficient of determination (r^2) = .7744, which means that 77.4% of the variance in cloud-computing technology can be explained by the reliability factor for government organizations.

There was a significant, strong, positive relationship between the reliability factor and the decision to adopt cloud computing in an academic organization, $r(35) = .86$, $p < .001$, two-tailed. The coefficient of determination (r^2) = .7396, which means that 74% of the variance in cloud-computing technology can be explained by the reliability factor in academic organizations.

There was a significant, strong, positive relationship between the reliability factor and the decision to adopt cloud computing in a corporate (for-profit) organization, $r(36) = .87$, $p < .001$, two-tailed. The coefficient of determination (r^2) = .7569, which means that 75.7% of the variance in cloud-computing technology can be explained by the reliability factor in corporate (for-profit) organizations. Scatterplots of these relationships are presented in appendix I.

$\mathbf{H_{03a}}$ stated that there is no correlation between the reliability factor and the decision to adopt cloud computing in a government organization. There was a significant, strong, positive relationship between the reliability factor and the decision to adopt cloud computing in a government organization, $r(33) = .88$, $p < .001$, two-tailed. Therefore, the null hypothesis was rejected.

$\mathbf{H_{03b}}$ stated that there is no correlation between the reliability factor and the decision to adopt cloud computing in a for-profit corporate organization. There was a significant, strong, positive relationship between the reliability factor and the decision to adopt cloud computing in a corporate (for-profit) organization, $r(36) = .87$, $p < .001$, two-tailed. Therefore, the null hypothesis was rejected.

$\mathbf{H_{03c}}$ stated that there is no correlation between the reliability factor and the decision to adopt cloud computing in an academic institution. There was a significant, strong, positive relationship between the reliability factor and the decision to adopt cloud computing in an academic organization, $r(35) = .86, p < .001$, two-tailed. Therefore, the null hypothesis was rejected.

What is the difference between organizations on the decision to adopt cloud computing? Research question 4 was invested with a one-way ANOVA. The independent variable was organization type with three levels (government, academic, and corporate/for-profit). The dependent variable was the decision to adopt cloud-computing technology. Group means for organization type by decision to adopt cloud computing are presented in table 13.

Table 13. Group Means for Organization Type by Decision to Adopt Cloud Computing Technology

Variable	*n*	*M*	*SD*
Government	35	4.07	0.47
Academic	37	4.05	0.43
Corporate (For-profit)	38	4.33	0.40
Total	110	4.15	0.45

Levene's Test for Homogeneity of Variance indicated that the assumption had not been violated, $p = .560$. See table 14.

Table 14. Levene's Test for Homogeneity of Variance

Levene Statistic	*df1*	*df2*	*p*
.583	2	107	.560

There was a significant difference between organizations relative to the decision to adopt cloud computing, $F(2, 107) = 4.66, p = .011$. The ANOVA summary table is presented in table 15.

Table 15. ANOVA Summary Table

Source	Sum of Squares	*df*	Mean Square	*F*	*p*
Between Groups	1.75	2	0.87	4.66	.011
Within Groups	20.09	107	0.19		
Total	21.84	109			

Scheffe post hoc comparisons were conducted on the data to determine where the significant differences existed. The results are presented in table 16.

Table 16. Scheffe Post Hoc Comparisons

	(I) Organization Type	(J) Organization Type	Mean Difference (I-J)	Std. Error	*p*
Scheffe	Government	Academic	.02	.10	.978
		Corporate (For-profit)	-.25*	.10	.048
	Academic	Government	-.02	.10	.978
		Corporate (For-profit)	-.27*	.10	.026
	Corporate (For-profit)	Government	.25*	.10	.048
		Academic	.27*	.10	.026

The decision to adopt cloud-computing technology was significantly higher for corporate (for-profit) organizations) than for government (p = .048) and academic (p = .026) organizations. However, there was no significant difference in the decision to adopt cloud-computing technology between government and academic organizations (p = .978). See figure 15.

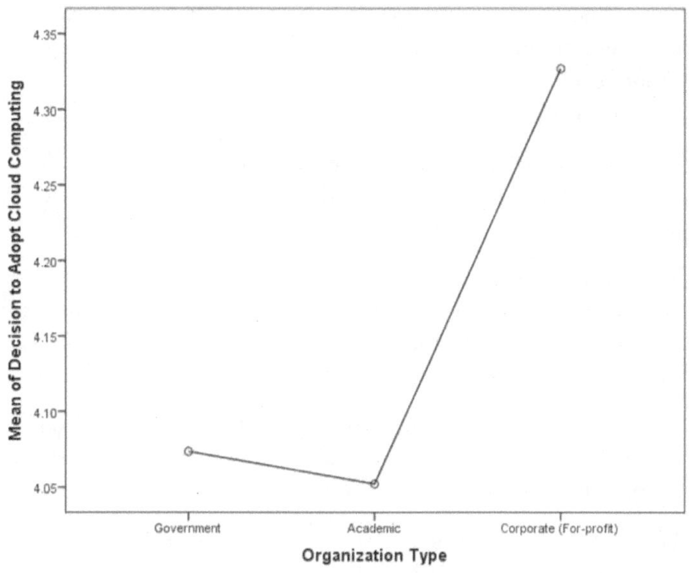

Figure 15. Decision to Adopt Cloud Computing by Organization Type

H_{04} stated that there is no difference between adoption decisions between a government organization, a for-profit corporate organization, and an academic institution. There was a significant difference between organizations relative to the decision to adopt cloud computing, $F(2, 107) = 4.66$, $p = .011$. Therefore, the null hypothesis was rejected. Hypotheses and outcomes are summarized in table 17.

Table 17. Hypotheses and Outcomes

Hypothesis	Significance	Outcome
H_{01a}: There is no correlation between the security effectiveness factor and the decision to adopt cloud computing in a government organization.	$p < .001$	Null Rejected.
H_{01b}: There is no correlation between the security effectiveness factor and the decision to adopt cloud computing in a for-profit corporate organization.	$p < .001$	Null Rejected.

H_{01c}: There is no correlation between the security effectiveness factor and the decision to adopt cloud computing in an academic institution.	$p < .001$	Null Rejected.
H_{02a}: There is no correlation between the cost-savings factor and the decision to adopt cloud computing in a government organization.	$p < .001$	Null Rejected.
H_{02b}: There is no correlation between the cost-savings factor and the decision to adopt cloud computing in a for-profit corporate organization.	$p < .001$	Null Rejected.
H_{02c}: There is no correlation between the cost-savings factor and the decision to adopt cloud computing in an academic institution.	$p < .001$	Null Rejected.
H_{03a}: There is no correlation between the reliability factor and the decision to adopt cloud computing in a government organization.	$p < .001$	Null Rejected.
H_{03b}: There is no correlation between the reliability factor and the decision to adopt cloud computing in a for-profit corporate organization.	$p < .001$	Null Rejected.
H_{03c}: There is no correlation between the reliability factor and the decision to adopt cloud computing in an academic institution.	$p < .001$	Null Rejected.
H_{04}: There is no difference between adoption decisions between a government organization, a for-profit corporate organization, and an academic institution.	$p = .011$	Null Rejected

Results

Four research questions and ten associated hypotheses were formulated for dissertational inquiry. It was determined that there was a significant, strong, positive relationship between the security effectiveness factor and the decision to adopt cloud computing in a government organization. There was a significant, strong, positive relationship between the security effectiveness factor and the decision to adopt cloud computing in an academic organization. There was a significant, strong, positive relationship between the security effectiveness factor and the decision to adopt cloud computing in a corporate (for-profit) organization.

There was a significant, strong, positive relationship between the cost-savings factor and the decision to adopt cloud computing in a government organization. There was a significant, strong, positive relationship between the cost-savings factor and the decision to adopt cloud computing in an academic organization. There was a significant, strong, positive relationship between the cost-savings factor and the decision to adopt cloud computing in a corporate (for-profit) organization.

There was a significant, strong, positive relationship between the reliability factor and the decision to adopt cloud computing in a government organization. There was a significant, strong, positive relationship between the reliability factor and the decision to adopt cloud computing in an academic organization. There was a significant, strong, positive relationship between the reliability factor and the decision to adopt cloud computing in a corporate (for-profit) organization.

Lastly, there was a significant difference between organizations relative to the decision to adopt cloud computing. Specifically, the decision to adopt cloud computing was significantly higher for corporate (for-profit) organizations than for government and academic organizations. However, there was no significant difference in the decision to adopt cloud-computing technology between government and academic organizations. Implications and recommendations will be discussed in chapter 5.

Summary

The purpose of this correlational study was to discover the relationship between the key factors identified as cost savings, security, and reliability that influenced IT directors' decision to adopt cloud computing within their respective for-profit corporate entities, academic institutions, and government agencies in Central Florida. To collect data for this study, a preexisting previously validated questionnaire was adapted from Ali et al. (2016). The survey instrument was formatted using a 5-point Likert-type scale with answers ranging from strongly disagree to strongly agree. Reliability and validation of the survey instrument was conducted by Ali et al. (2016) in the form of a pilot study. The results of their pilot study revealed Cronbach alpha values of .80 or higher on each of the survey constructs. The research questions depicted in chapter 4 formed the foundation for this study to discover the relationship between the key factors that IT decision-makers focus on when they make their decision to adopt cloud computing.

The primary demographic sampled for this study consisted of 111 IT decision-makers from government (n = 35), academic (n = 38), and for-profit corporate (n = 38) organizations. Using SPSS analysis showed the largest group (n = 53) consisted of participants in the 35–44 age range with 59.9% of them being male. In terms of answering the research questions, correlational analysis revealed there was a statistically significant relationship between the security effectiveness factor, cost-savings factor, and the reliability factor and the decision to adopt cloud computing between a government organization, for-profit corporate organization, and an academic institution, thereby rejecting all the null hypotheses for the first three research questions. To answer the fourth research question, an ANOVA and Scheffe post hoc comparison test was conducted to determine the decision to adopt cloud computing was significantly higher for corporate (for-profit) organizations than for government and academic organizations. Chapter 5 provides the findings of this correlational study as well as recommendations for leadership and future research.

5

Conclusions and Recommendations

The purpose of this correlational study was to discover the relationship between the key factors (security effectiveness, cost savings, and reliability) that influence IT directors' decisions to adopt cloud-computing technology within their respective institutions. The key factors investigated in this study were reliability, cost, and security. A quantitative approach and correlational research design were used to determine which factor most influenced the decision to adopt cloud computing between government agencies, for-profit corporate organizations, and academic institutions in Central Florida. Although cloud computing is becoming widely adopted within various industries, previous research found many decision-makers did not understand the factors that contributed to the adoption of cloud computing (Gill 2011; Hus and Lin 2016; Low, Chen, and Wu 2011). To better understand which factors influenced the decision to adopt cloud computing, 111 IT decision-makers from government agencies, for-profit corporate organizations, and academic institutions were surveyed.

Survey responses were collected from participants across the Central Florida region, meeting the inclusion criteria to participate in the survey. Respondents varied in their age, gender, educational level, and role within their respective organization. While the demographic factors did not significantly contribute to the correlation, they did facilitate being able to generalize the study.

Research Questions/Hypotheses

RQ1: Is there a correlation between the security effectiveness factor and the decision to adopt cloud computing in a government organization, for-profit business organization, and an academic institution?

H_{01a}: There is no correlation between the security effectiveness factor and the decision to adopt cloud computing in a government organization.

H_{1a}: There is a correlation between the security effectiveness factor and the decision to adopt cloud computing in a government organization.

H_{01b}: There is no correlation between the security effectiveness factor and the decision to adopt cloud computing in a for-profit corporate organization.

H_{1b}: There is a correlation between the security effectiveness factor and the decision to adopt cloud computing in a for-profit corporate organization.

H_{01c}: There is no correlation between the security effectiveness factor and the decision to adopt cloud computing in an academic institution.

H_{1c}: There is a correlation between the security effectiveness factor and the decision to adopt cloud computing in an academic institution.

RQ2: Is there a correlation between the cost-savings factor and the decision to adopt cloud computing in a government organization, for-profit business organization, and an academic institution?

H_{02a}: There is no correlation between the cost-savings factor and the decision to adopt cloud computing in a government organization.

H_{2a}: There is a correlation between the cost-savings factor and the decision to adopt cloud computing in a government organization.

H_{02b}: There is no correlation between the cost-savings factor and the decision to adopt cloud computing in a for-profit corporate organization.

H_{2b}: There is a correlation between the cost-savings factor and the decision to adopt cloud computing in a for-profit corporate organization.

H_{02c}: There is no correlation between the cost-savings factor and the decision to adopt cloud computing in an academic institution.

H_{2c}: There is a correlation between the cost-savings factor and the decision to adopt cloud computing in an academic institution.

RQ3: Is there a correlation between the reliability factor and the decision to adopt cloud computing in a government organization, for-profit business organization, and an academic institution?

H_{03a}: There is no correlation between the reliability factor and the decision to adopt cloud computing in a government organization.

H_{3a}: There is a correlation between the reliability factor and the decision to adopt cloud computing in a government organization.

H_{03b}: There is no correlation between the reliability factor and the decision to adopt cloud computing in a for-profit corporate organization.

H_{3b}: There is a correlation between the reliability factor and the decision to adopt cloud computing in a for-profit corporate organization.

H_{03c}: There is no correlation between the reliability factor and the decision to adopt cloud computing in an academic institution.

H_{3c}: There is a correlation between the reliability factor and the decision to adopt cloud computing in an academic institution.

RQ4: What is the difference between organizations on the decision to adopt cloud computing?

$H_{4:}$ There is a difference between adoption decisions between a government organization, a for-profit corporate organization, and an academic institution.

H_{04}: There is no difference between adoption decisions between a government organization, a for-profit corporate organization, and an academic institution.

Discussion of Findings

Data analysis for this study utilized IBM SPSS software and incorporated descriptive statistics, Pearson r correlation, and analysis of variance to determine statistical significance. The independent variables used in this study were security effectiveness, cost savings, and reliability with the adoption decision as the dependent variable. There were four research questions that guided this study. To begin answering the research questions, the responses to the survey were divided into three subgroups correlating to the organizational type (government agency, academic institution, and for-profit corporate organization) of the respondent. Pearson r correlation was accomplished against the data to measure the strength of the linear association between the variables.

For research question 1, is there a correlation between the security effectiveness factor and the decision to adopt cloud computing in a government organization, for-profit business organization, and an academic institution? Based on the analysis, there was a statistically significant relationship between the security effectiveness factor and the decision to adopt cloud computing in all three organizational types with the following results: government agency $r(33) = .83$, $p < .001$, two-tailed; academic institution $r(35) = .86$, $p < .001$, two-tailed; and for-profit corporate organization $r(36) = .74$, $p < .001$ two-tailed, thereby rejecting all the null hypotheses for question 1.

Question 2 asked if there was a correlation between the cost-savings factor and the decision to adopt cloud computing in a government organization, for-profit business organization, and an academic institution. Correlational analysis of the data revealed there was a statistically positive relationship between cost savings and the decision to adopt in all three organizational types with the following results: government $r(33) = .90$, $p < .001$, two-tailed; academic institution $(35) = .89$, $p <.001$; and for-profit organization $r(36) = .93$, $p < .001$, thereby rejecting all the null hypotheses for question 2.

Question 3 asked if there was a correlation between the reliability factor and the decision to adopt cloud computing in a government organization, for-profit business organization, and an academic institution. Correlational analysis of the data revealed there was a statistically positive relationship between the reliability factor and the decision to adopt in all three organizational types with the following results: government $r(33) = .88$, $p < .001$, two-tailed; academic institution $(35) = .86$, $p < .001$; and for-profit organization $r(36) = .87$, $p < .001$, thereby rejecting all the null hypotheses for question 3.

The fourth and final research question asked if there was a difference between organizations on the decision to adopt cloud computing. To answer this question, a one-way analysis of variance (ANOVA) had to be conducted to determine if there was a statistical difference between the groups when deciding to adopt cloud computing. To conduct the ANOVA, the organizational type (government, academic, and corporate/for-profit) was determined to be the independent variable with the decision to adopt being the dependent variable. Using the Levene test for homogeneity of variance, it was revealed the assumption that the variance of populations was equal. The results of the one-way ANOVA revealed there was a statistically significant difference between the groups; however, to determine which specific group was statistically different from the others, a post hoc test was required.

A Scheffe post hoc comparison test was conducted to determine which group was statistically different. Results of the testing revealed the decision to adopt cloud-computing technology was significantly higher for corporate (for-profit) organizations than for government ($p = .048$) and academic ($p = .026$) organizations.

Limitations

The purpose of this correlational study was to discover the relationship between the key factors (security effectiveness, cost savings, and reliability) that influence IT directors' decisions to adopt cloud-computing technology within their respective institutions. A limitation of this study was the sampled population, which was restricted to participants from either a government agency, a for-profit corporate organization, or an academic institution of higher learning. Another potential limitation of this study was generalizing the results to the adoption of other IT-related technologies. As this study focused on the adoption of cloud computing, the findings could not necessarily be applied broadly to other technologies. Finally, this study only focused on participants within the Central Florida region of the US.

Recommendations to Leaders and Practitioners

The analysis indicated decision-makers of for-profit corporate organizations have less aversion to risk for adopting new technologies. Cloud computing has become the new trend transforming the technology landscape across various industries. While leaders in corporate organizations have shown to be early adopters of cloud-computing technologies, it is recommended IT decision-makers use the results of this study to continue to mature their cloud-computing strategies and invest in multiple cloud implementation models (private, public, or hybrid) in order to keep pace with or maintain an advantage over their competitors in the marketplace. Cloud computing has caused a paradigm shift in the field of IT that has influenced the development of new business strategies, models, and processes. As such, practitioners working in the corporate industry will need to expand their skillsets to include cloud capabilities to facilitate the implementation of governance and compliance programs as they work with cloud service providers.

Leaders in government agencies can use the results to aid in the deployment of cloud-computing services across government agencies to reduce the footprint of the large-scale IT implementation traditionally seen within the government. Additionally, the results can be used as a benchmark for articulating security requirements for cloud implementations within the classified government environments. Understanding the limitations of security with cloud implementations, government leaders can use the results of this study to establish minimum security baselines, while practitioners within the government sector will need to change their perspective as it relates to security in the cloud environment, focusing more on the end-point devices that have direct connections to the cloud environment than the infrastructure components. Practitioners are used to developing security controls for on-premise hardware and software; however, with cloud implementations, the cloud service provider is responsible for the security of the cloud environment.

While the adoption of cloud computing in academic institutions lagged behind corporate organizations and government agencies, research revealed academia was able to leverage several benefits from implementing cloud computing technologies. With current world conditions dealing with COVID-19 and having to adhere to social distancing, more and more academic institutions are having to rely on cloud-computing technologies such as PaaS and SaaS to deliver the virtual classroom. In order to attract new students, leaders of academic institutions will need to redesign their IT operations to be aligned with the various cloud-computing technologies as well as modernize how education services are presented to keep pace with the major paradigm shift that has occurred in order to reduce costs and remain competitive with other academic institutions of higher learning.

Recommendations for Future Research

Given the results of this study, it is widely understood that 95 percent of government, academic, and for-profit corporate organizations in Central Florida have already adopted some variation of cloud computing. While this study focused on the factors of cost savings, reliability, and security, it did not specifically address which cloud implementation (private, public, community or hybrid) was being used or was preferred by the sampled organizations. Additionally, this study did not capture if other factors not addressed as part of this research study were significant in the decision to adopt cloud computing. This study was very specific in the targeted population. Given the sensitive surrounding medical and financial data, it is recommend a similar quantitative or qualitative study be done specifically targeting the medical and financial industries to assess their appetite for moving to a complete cloud-based solution for managing their records.

Another potential area for future study would be to consider the effect of organizational size as an influence on the decision to adopt cloud computing. Larger organizations may view the relevant factors contributing to the adoption of cloud computing differently than smaller organizations. Finally, another option for future research would be to conduct this study outside the United States using the same targeted population.

Summary

The purpose of this correlational study was to discover the relationship between the key factors (security effectiveness, cost savings, and reliability) that influence IT directors' decisions to adopt cloud-computing technology within their respective institutions. The research questions that guided this study were the following:

RQ1: Is there a correlation between the security effectiveness factor and the decision to adopt cloud computing in a government organization, for-profit business organization, and an academic institution?

RQ2: Is there a correlation between the cost-savings factor and the decision to adopt cloud computing in a government organization, for-profit business organization, and an academic institution?

RQ3: Is there a correlation between the reliability factor and the decision to adopt cloud computing in a government organization, for-profit business organization, and an academic institution?

RQ4: What is the difference between organizations on the decision to adopt cloud computing?

The results of the study found there was a statistically significant correlation between the security effectiveness, cost savings, and reliability factors and the decision to adopt cloud computing in government organizations, for-profit corporate organizations, and academic institutions. However, between the organizations, the study revealed the decision to adopt cloud computing was significantly higher in for-profit corporate organizations than for government or academic institutions. A key implication for leaders was while cloud computing has been adopted by 95 percent of the organizations sampled coupled with the onset of the COVID-19 pandemic, a greater reliance on cloud-computing technologies has been the result. As such,

IT decision-makers should realize the importance of establishing governance and compliance programs specifically targeted at cloud implementations while continuing to mature their cloud-computing strategies to maintain a competitive advantage in the marketplace.

APPENDICES

APPENDIX A

G* Power Output

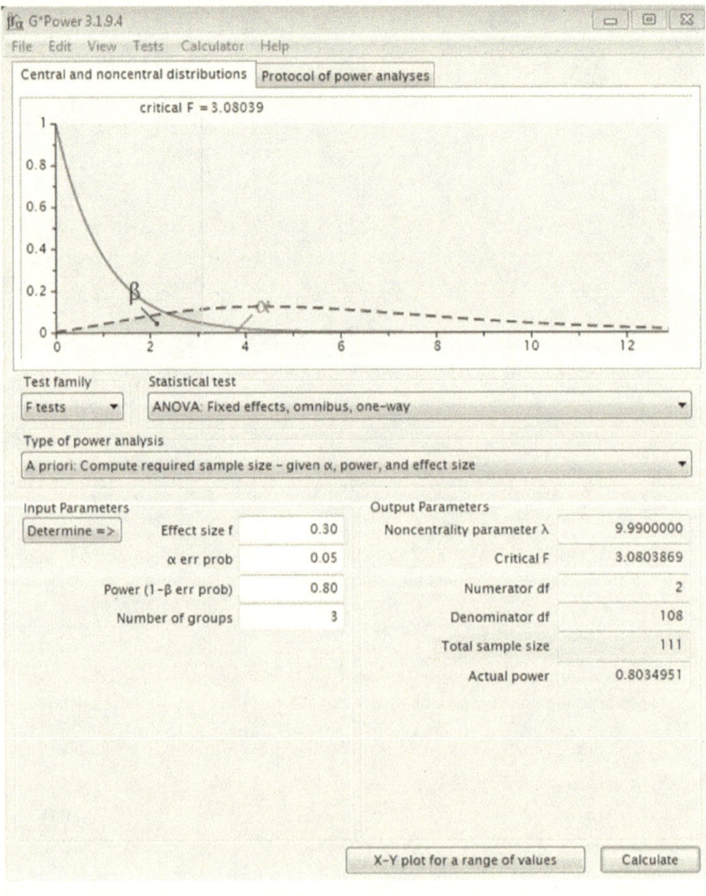

APPENDIX B

Survey Instrument

Questionnaire Instructions

This survey contains basic demographic questions followed by 3 primary questions supported by 5 sub-statements each. Answering this questionnaire should not take more than 8–10 minutes.

The purpose of this quantitative nonexperimental, correlational survey study is to discover the relationship between the key factors that influence IT decision-maker's decisions to adopt cloud-computing technology within their respective for-profit corporate entities, academic institutions, and government agencies in Central Florida. Demographic data is collected to aid in the analysis and decomposition of survey response data to better understand certain background characteristics of meaningful groups of respondents.

Section one
1. **What is your age? (Demographic Question)**
☐ Under 18 ☐ 18–24 ☐ 25–34 ☐ 35–44 ☐ 45–54 ☐ 55–64 ☐ 65+
2. **What is your gender? (Demographic Question)**
☐ Male ☐ Female
3. **What is your educational level? (Demographic Question)**
☐ Undergraduate ☐ Graduate ☐ Post graduate ☐ Other
4. **What is your role? (Demographic Question)**
☐ CIO Dep ☐ CIO ☐ CTO Dep ☐ CTO ☐ IT Director ☐ Equivalent level position ☐ Other
5. **What is your organization's type? (Demographic Question)**
☐ Government ☐ Academic ☐ Corporate (For-profit)
6. **Is your organization within Central Florida? (Demographic Question)**
☐ Yes ☐ No
7. **Does your organization use cloud computing? (Demographic Question)**
☐ Yes ☐ No
8. **If Yes, what type of cloud delivery model does your organization use (Check all that apply)?** ☐ Software-as-a-Service ☐ Platform-as-a-Service ☐ Infrastructure-as-a-Service ☐ Other

9. **What cloud-based services/applications are used by your organization? (Check all that apply)**

- ☐ Email
- ☐ SMS/Text messaging
- ☐ Telephone Services/VoIP
- ☐ Office Applications
- ☐ Web Conferencing
- ☐ Customer Relationship Management (CRM)
- ☐ Virtual Computers/Servers
- ☐ Data Backup/Storage
- ☐ Disaster Recovery
- ☐ Website Hosting
- ☐ Operating System
- ☐ E-commerce
- ☐ Remote Access/VPN
- ☐ Social Networking
- ☐ Payroll
- ☐ Billing and invoicing

10. Are IT decision-makers influenced by a security factor when they decide to adopt cloud computing?		Strongly Disagree	Disagree	Neither Agree or Disagree	Agree	Strongly Agree
1	Data security is a major concern for organizations considering to adopt cloud computing.					
2	Cloud computing provides a sufficient security transfer channel during the process of mass data interchange.					
3	Loss of control over data and applications influences the adoption decision.					
4	Cloud computing provides a secure service.					
5	Security concerns are not an issue with cloud computing.					

11. Are IT decision-makers influenced by a cost factor when they decide whether to adopt cloud computing?		Strongly Disagree	Disagree	Neither Agree or Disagree	Agree	Strongly Agree
1	Cloud computing decreases the investment cost in new IT infrastructure.					
2	Cloud computing is more cost-effective compared with other IS technologies.					
3	Cloud computing reduces the costs of system upgrades.					
4	Cloud computing may increase the ability of an organization to adapt rapidly and cost efficiently in response to changes in the government, academic, or business environment.					
5	Cloud computing reduces the total cost of operational processes.					

	12. Are IT decision-makers influenced by a reliability factor when they decide to adopt cloud computing technology?	Strongly Disagree	Disagree	Neither Agree or Disagree	Agree	Strongly Agree
1	Cloud computing provides a reliable service with high availability.					
2	Cloud computing can/may enable better communication between our suppliers and customers.					
3	Cloud computing can create a flexible environment to operate in.					
4	Cloud computing can provide aces to cloud services from various client devices (e.g., thick clients, thin clients, laptops, tablets, smartphones, etc.) in our environment.					
5	Cloud computing can enable the organization to respond quickly to customer requests.					

APPENDIX C

Informed Consent

INFORMED CONSENT: PARTICIPANTS 18 YEARS OF AGE AND OLDER

Greetings,

My name is Sean Lawrence and I am a doctoral student with the University of Phoenix doing my research for the Doctorate in Management, with a concentration in Information Systems Technology. I am conducting a research study entitled Factors Influencing the Adoption of Cloud Computing between Government Agencies, Academic Institutions, and For-Profit Corporate Organizations in Central Florida.

The purpose of the research study is to discover the relationship between the key factors that influence the IT Director's decisions to adopt cloud-computing technology within their respective for-profit corporate entities, academic institutions, and government agencies in Central Florida.

To participate in this study, you must be a principal decision-maker within your organization's Information Technology (IT) department that has significant influence over IT procurement decisions. Your participation in this study will involve answering basic demographic questions followed by completing a survey consisting of 15 questions. The survey should only take 8–10 minutes to complete. Only responses from participants meeting the inclusion criteria outlined above will be included.

You can decide to be a part of this study or not. Once you start, you can withdraw from the study at any time without any repercussions. The results of the research study may be published but your identity will remain…anonymous and your name will not be made known to any outside parties.

In this research, there are no foreseeable risks to you.

Although there may be no direct benefit to you, a possible benefit from your being part of this study is I hope to learn which factors influence IT Directors or equivalents to make the decision to adopt cloud computing technology in their respective organizations in Central Florida. The study could help IT Directors, CIO's and other decision-makers to make an informed decision about cloud computing adoption in Central Florida.

If you have any questions about the research study, please call me at or email me at . edu. For questions about your rights as a study participant, or any concerns or complaints, please contact the University of Phoenix Institutional Review Board at IRB@phoenix.edu.

As a participant in this study, you should understand the following:

1. You may decide not to be part of this study or you may want to withdraw from the study at any time. Should you choose to withdraw, please be advised because of the anonymous nature of this study any data you input will not be able to be identified for removal. If you want to withdraw, please call me at or email me at .
2. Your identity will be kept anonymous.
3. Sean Lawrence, the researcher, has fully explained the nature of the research study and has answered all of your questions and concerns.

Although, some survey data will initially reside on Survey Monkey's servers for initial data analysis, the aggregated raw data will be downloaded into an Excel spreadsheet on the researchers' computer and deleted from Survey Monkey's website. To ensure safe keeping of participant responses all survey data will be stored in a password protected folder on the researcher's computer. The researcher will be the only person with the encrypted password to access the data. Upon completion of the study and acceptance of the results by the university, all associated electronic files will continue to be maintained for a period of three years following approval of the dissertation in accordance with university research retention polices. Once the research files have reached the end of life per university polices for retention, all files associated with this study will be destroyed using a file-shredding software application that deletes the contents of the file and completely overwrites the data to ensure it cannot be recovered.

4. The results of this study may be published.

By accepting this agreement, you agree that you understand the nature of the study, the possible risks and benefits to you as a participant, and how your identity will be kept anonymous. When selecting "Yes," this means that you are 18 years of age or older and that you give your permission to volunteer as a participant in the study that is described here and will be taken to the survey to complete. If "No" is selected, you will exit the survey.

(☐) Yes, I accept the above terms.
(☐) No, I do not accept the above terms.

APPENDIX D

Request to Use Survey Instrument

Good Day, Mr. Ali

Reference your journal article: Factors to be considered in cloud computing (2016)

My name is Sean Lawrence and I am a doctoral student with the University of Phoenix doing my research for the Doctorate in Management, with a concentration in Information Systems Technology.

My dissertation proposal focuses on the Factors Influencing the Adoption of Cloud Computing between Government Agencies, Academic Institutions, and for-profit corporate organizations in Central Florida.

I have read your journal article and found it is similar to my topic and believe the survey instrument you reference in your article may prove valuable within my own research.

I, therefore, would like to request a copy of your survey instrument to review and see if it would fit within my dissertation research.

If it turns out to be a compatible instrument, I would follow up and request your permission to use, modify, and adapt your survey instrument for use within my study.

Your favorable consideration would be greatly appreciated.
I look forward to hearing from you.
Thank you very much.

Respectfully,
Sean C. Lawrence

APPENDIX E

Approval to Use Survey Instrument

Re: Request a copy of survey instrument (Follow-up)
Omar Ali <Omar.Ali@usq.edu.au>
Sat 1/11/2020 4:24 PM
To: Sean Lawrence <seanlawrence@email.phoenix.edu>

Dear Sean,

You can use this email as a permission from me to use, modify, and adopt my research instrument survey tools in your research study.

Good luck,
Omar Ali
Get Outlook for iOS

From: Sean Lawrence <seanlawrence@email.phoenix.edu>
Sent: Saturday, January 11, 2020 11:31 pm
To: Omar Ali
Cc: Douglas.Neeley
Subject: Re: Request a copy of survey instrument (Follow-up)

Dr. Ali

Sir just wanted to follow up on my official request below.

Respectfully,
Sean Lawrence

From: Sean Lawrence <seanlawrence@email.phoenix.edu>
Sent: Tuesday, January 7, 2020 10:24:46 PM
To: Omar Ali <Omar.Ali@usq.edu.au>
Cc: Douglas.Neeley <Douglas.Neeley@Phoenix.edu>
Subject: Re: Request a copy of survey instrument

Dr. Ali,

Thank you again for sending the complete survey for me to review.

After reviewing the survey, I have determined it is a good fit for my research study. I therefore officially request your permission to use, modify and adapt your survey instrument for use within my study.

I will ensure credit for the survey instrument is properly given to you and your coauthors within my dissertation.

Your favorable consideration would be greatly appreciated.

Respectfully,
Sean C. Lawrence
UoPX Doctoral Candidate

From: Omar Ali <Omar.Ali@usq.edu.au>
Sent: Tuesday, January 7, 2020 12:35 AM
To: Sean Lawrence <seanlawrence@email.phoenix.edu>
Subject: Re: Request a copy of survey instrument

Dear Sean,

Could you please check the attached file. Also, feel free to contact me at any me if you need any help. Good luck,

Omar Ali
Ph.D. Management Information Systems
School of Management and Enterprise
Faculty of Business, Education, Law, and Art
University of Southern Queensland
Office: +61 7 4631 5387
Mobile: +61 411 593 870
E-mail: Omar.Ali@usq.edu.au

From: Sean Lawrence <seanlawrence@email.phoenix.edu>
Sent: Tuesday, 7 January 2020 2:54 PM
To: Omar Ali <Omar.Ali@usq.edu.au>
Subject: Request a copy of survey instrument

Good Day, Mr. Ali

Reference your journal article: Factors to be considered in cloud computing (2016)

My name is Sean Lawrence and I am a doctoral student with the University of Phoenix doing my research for the Doctorate in Management, with a concentration in Information Systems Technology.

My dissertation proposal focuses on the Factors Influencing the Adoption of Cloud Computing between Government Agencies, Academic Institutions, and for-profit corporate organizations in Central Florida.

I have read your journal article and found it is similar to my topic and believe the survey instrument you reference in your article may prove valuable within my own research.

I, therefore, would like to request a copy of your survey instrument to review and see if it would fit within my dissertation research.

If it turns out to be a compatible instrument, I would follow up and request your permission to use, modify, and adapt your survey instrument for use within my study.

Your favorable consideration would be greatly appreciated.

I look forward to hearing from you.

Thank you very much.
Respectfully,
Sean C. Lawrence

APPENDIX F

Histograms for Variables after Removal of Outlier

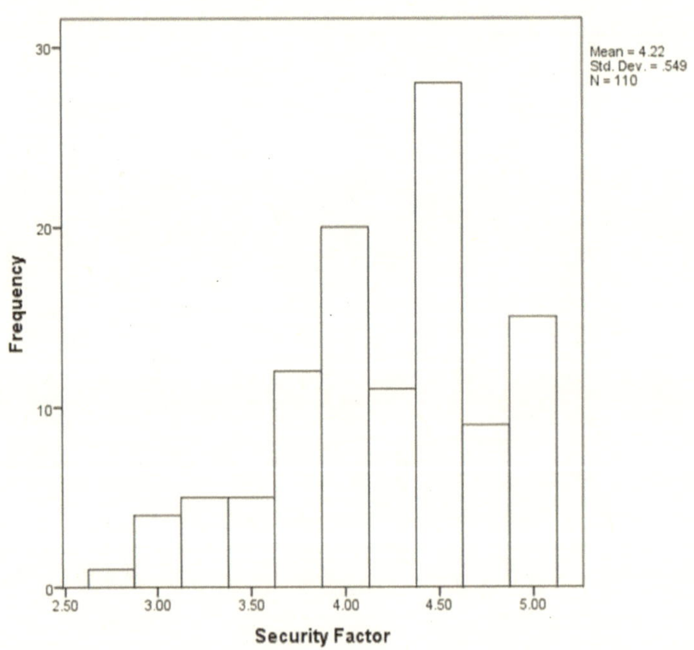

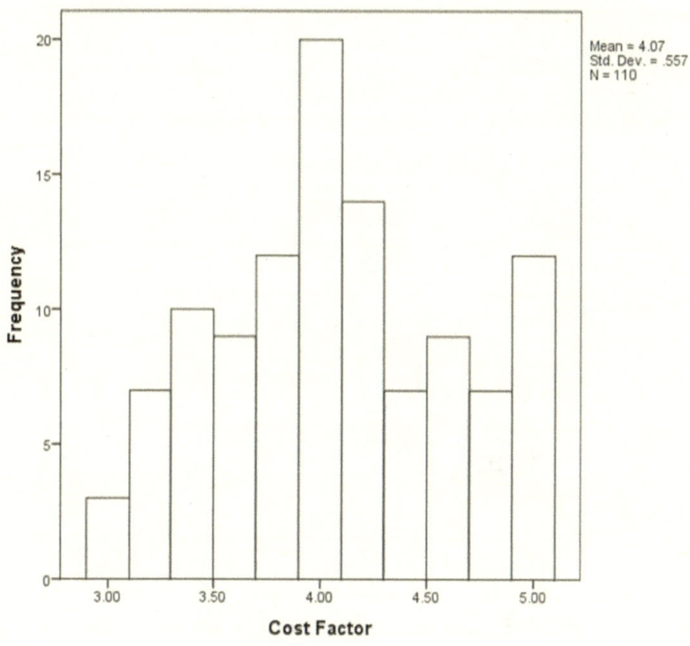
Mean = 4.07
Std. Dev. = .557
N = 110
Frequency
20
15
10
5
0
3.00
3.50
4.00
4.50
5.00
Cost Factor

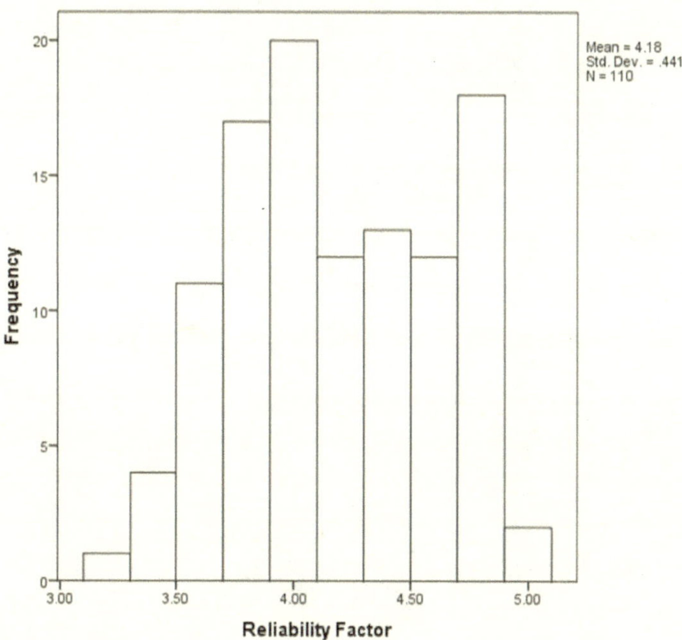
Mean = 4.18
Std. Dev. = .441
N = 110
Frequency
20
15
10
5
0
3.00
3.50
4.00
4.50
5.00
Reliability Factor

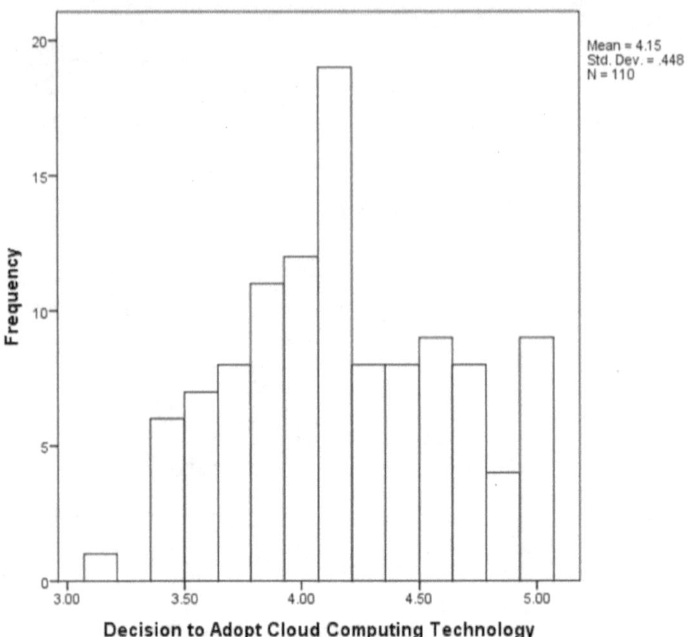
Mean = 4.15
Std. Dev. = .448
N = 110
Frequency
0
5
10
15
20
3.00
3.50
4.00
4.50
5.00
Decision to Adopt Cloud Computing Technology

APPENDIX G

Correlation Matrix

Correlations

Organization Type			Decision to Adopt	Security Factor	Cost Factor	Reliability Factor
Government	Decision to Adopt	Pearson Correlation	1	.834**	.903**	.884**
		Sig. (2-tailed)		.000	.000	.000
		N	35	35	35	35
	Security Factor	Pearson Correlation	.834**	1	.600**	.602**
		Sig. (2-tailed)	.000		.000	.000
		N	35	35	35	35
	Cost Factor	Pearson Correlation	.903**	.600**	1	.733**
		Sig. (2-tailed)	.000	.000		.000
		N	35	35	35	35
	Reliability Factor	Pearson Correlation	.884**	.602**	.733**	1
		Sig. (2-tailed)	.000	.000	.000	
		N	35	35	35	35
Academic	Decision to Adopt	Pearson Correlation	1	.836**	.885**	.856**
		Sig. (2-tailed)		.000	.000	.000
		N	37	37	37	37
	Security Factor	Pearson Correlation	.836**	1	.611**	.574**
		Sig. (2-tailed)	.000		.000	.000
		N	37	37	37	37
	Cost Factor	Pearson Correlation	.885**	.611**	1	.640**
		Sig. (2-tailed)	.000	.000		.000
		N	37	37	37	37
	Reliability Factor	Pearson Correlation	.856**	.574**	.640**	1
		Sig. (2-tailed)	.000	.000	.000	
		N	37	37	37	37
Corporate (For-profit)	Decision to Adopt	Pearson Correlation	1	.744**	.932**	.873**
		Sig. (2-tailed)		.000	.000	.000
		N	38	38	38	38
	Security Factor	Pearson Correlation	.744**	1	.544**	.445**
		Sig. (2-tailed)	.000		.000	.005
		N	38	38	38	38
	Cost Factor	Pearson Correlation	.932**	.544**	1	.774**
		Sig. (2-tailed)	.000	.000		.000
		N	38	38	38	38
	Reliability Factor	Pearson Correlation	.873**	.445**	.774**	1
		Sig. (2-tailed)	.000	.005	.000	
		N	38	38	38	38

**. Correlation is significant at the 0.01 level (2-tailed).

APPENDIX H

Scatterplots of Security Effectiveness and the Decision to Adopt Cloud-Computing Technology by Organization Type

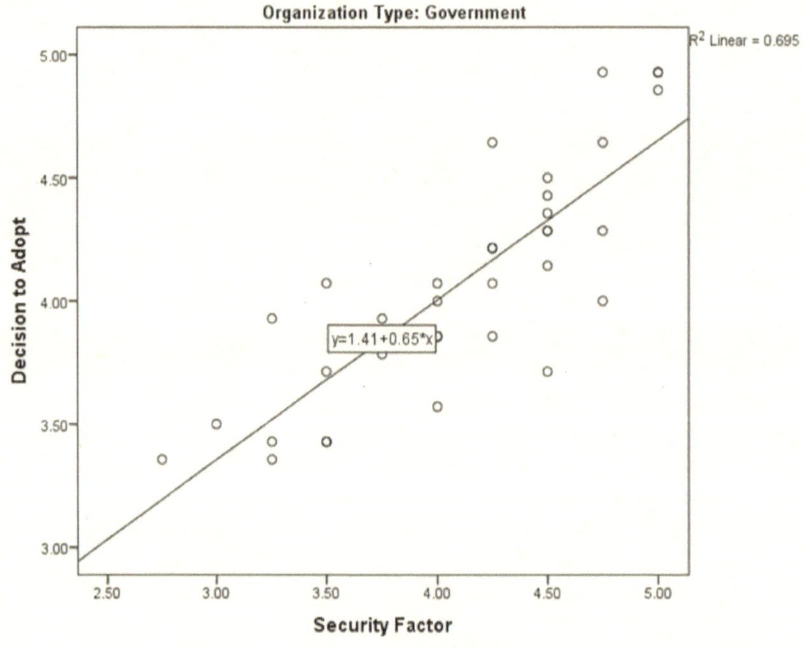

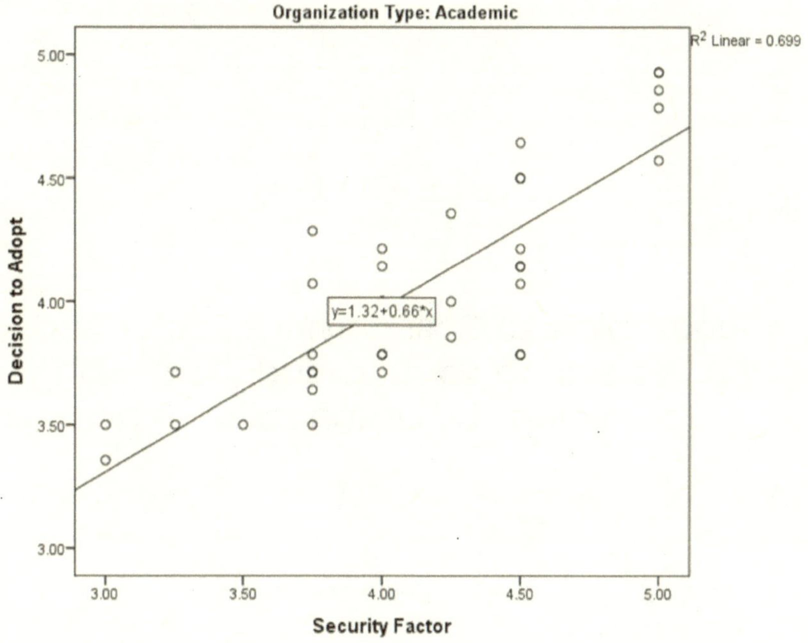
Organization Type: Academic
R² Linear = 0.699
y=1.32+0.66*x
Decision to Adopt
Security Factor
5.00
4.50
4.00
3.50
3.00
3.00
3.50
4.00
4.50
5.00

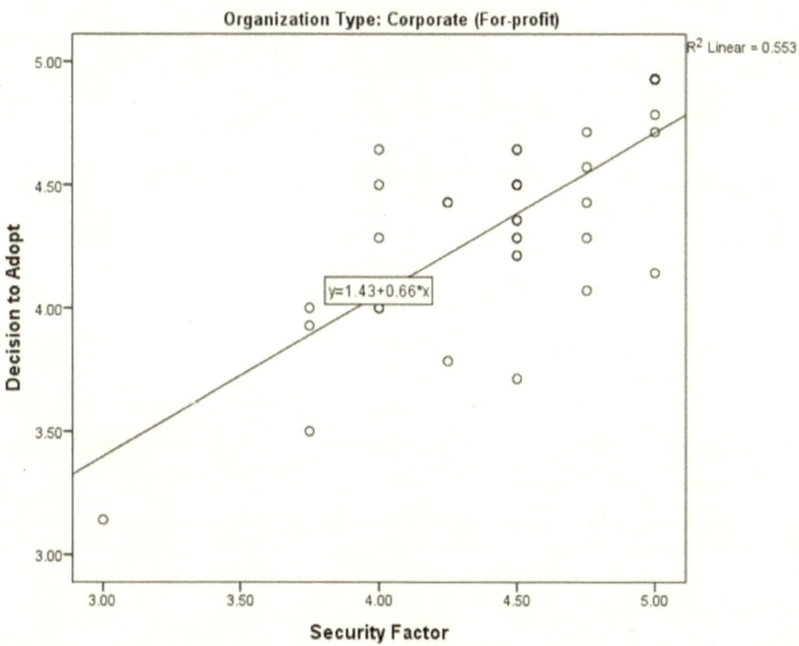
Organization Type: Corporate (For-profit)
R² Linear = 0.553
y=1.43+0.66*x
Decision to Adopt
Security Factor
5.00
4.50
4.00
3.50
3.00
3.00
3.50
4.00
4.50
5.00

APPENDIX I

Scatterplots of Cost-Savings Factor and the Decision to Adopt Cloud-Computing Technology by Organization Type

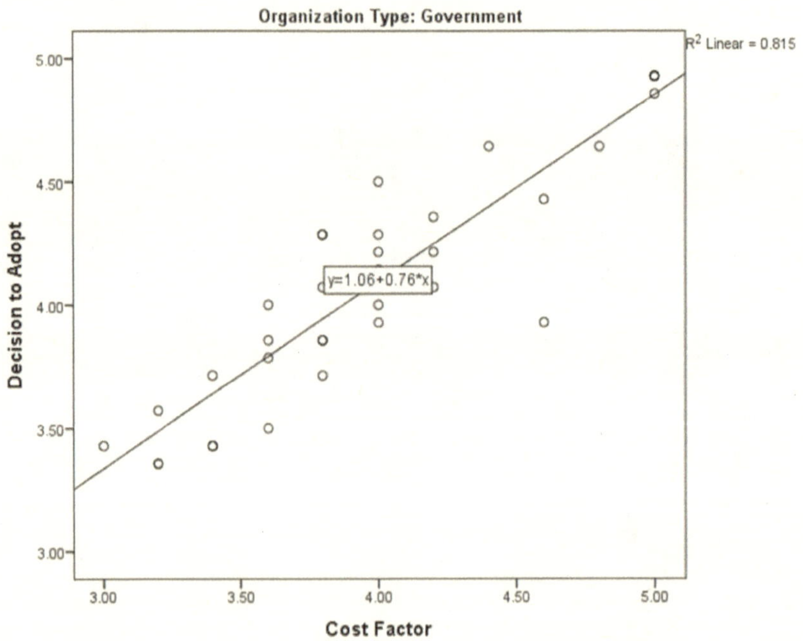

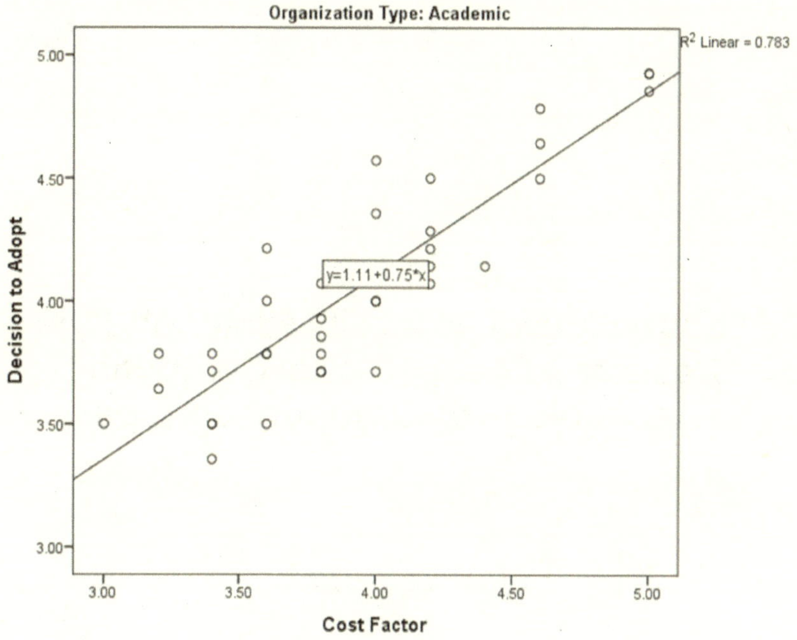
Organization Type: Academic
R² Linear = 0.783
y=1.11+0.75*x
Decision to Adopt
Cost Factor
5.00
4.50
4.00
3.50
3.00
3.00
3.50
4.00
4.50
5.00

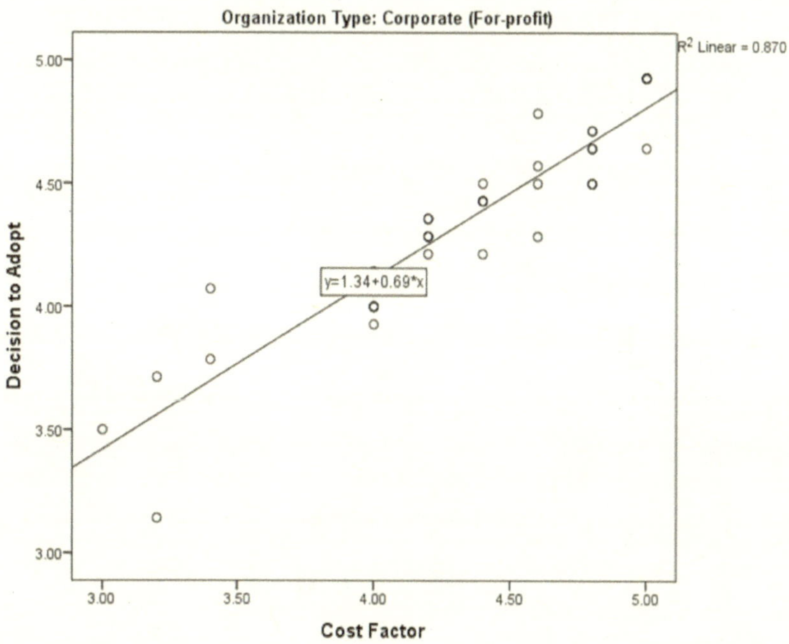
Organization Type: Corporate (For-profit)
R² Linear = 0.870
y=1.34+0.69*x
Decision to Adopt
Cost Factor
5.00
4.50
4.00
3.50
3.00
3.00
3.50
4.00
4.50
5.00

APPENDIX J

Scatterplots of Reliability Factor and the Decision to Adopt Cloud-Computing Technology by Organization Type

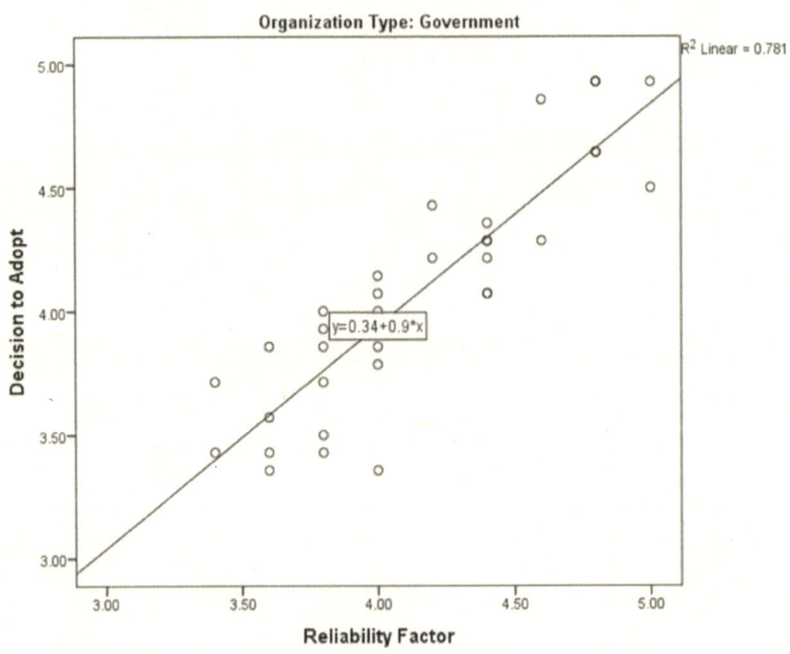

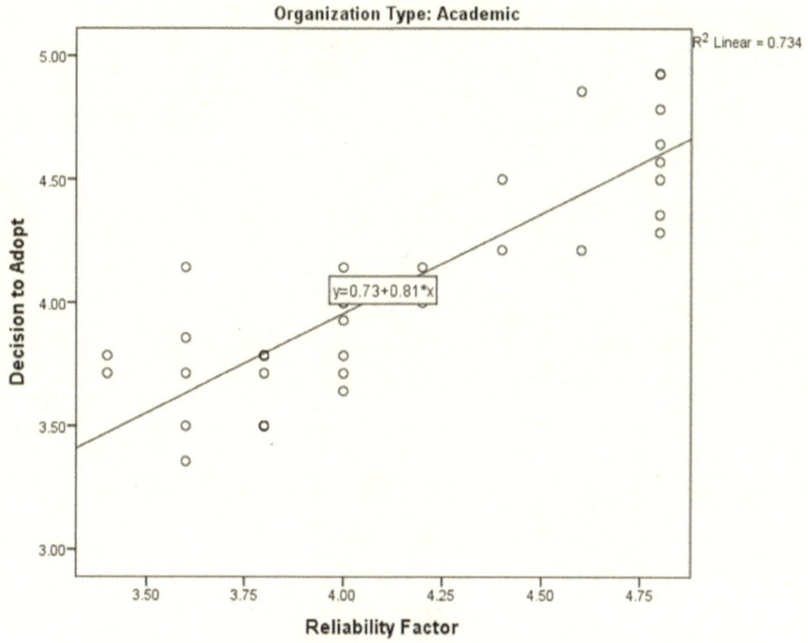
Organization Type: Academic
R² Linear = 0.734
y=0.73+0.81*x
Decision to Adopt
5.00
4.50
4.00
3.50
3.00
3.50
3.75
4.00
4.25
4.50
4.75
Reliability Factor

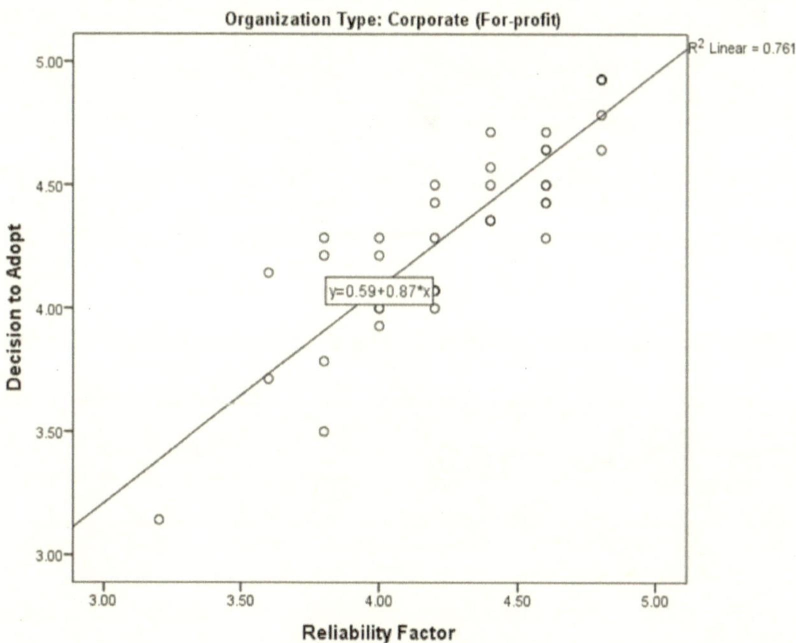
Organization Type: Corporate (For-profit)
R² Linear = 0.761
y=0.59+0.87*x
Decision to Adopt
5.00
4.50
4.00
3.50
3.00
3.00
3.50
4.00
4.50
5.00
Reliability Factor

REFERENCES

Aguinis, H., N. S. Hill, and J. R. Bailey. 2019. "Best Practices in Data Collection and Preparation: Recommendations for Reviewers, Editors, and Authors." *Organizational Research Methods 1*. 1–16. https://doi.org/10.1177/1094428119836485.

Ajzen, I. 1985. "From Intentions to Actions: A Theory of Planned Behavior." In *Action Control: From Cognition to Behavior*, edited by J. Kuhl and J Beckmann, 11–39. Berlin: Springer Berlin, Heidelberg.

———. 1991. "The Theory of Planned Behavior." *Organizational Behavior and Human Decision Processes* 50, no. 2 (December): 179–211.

Ajzen, I., and M. Fishbein. 1980. *Understanding Attitudes and Predicting Social Behavior*. Englewood Cliffs, NJ: Prentice-Hall.

Ajzen, I., and T. J. Madden. 1986. "Prediction of Goal-Directed Behavior: Attitudes, Intentions, and Perceived Behavioral Control." *Journal of Experimental Social Psychology* 22, no. 5: 453–474.

Akintomide, O. A. 2013. "Cloud Computing: The Third Revolution in IT." *Library of Progress Library Science, Information Technology & Computer* 33, no. 1 (December): 77–95.

Al-Ruithe, M., and E. Benkhelifa. 2017. "Analysis and Classification of Barriers and Critical Success Factors for Implementing a Cloud Data Governance Strategy." *Procedia Computer Science* 113: 223–232.

Alfifi, F. 2015. "Factors Influencing the Adoption of Cloud Computing by CIOs and Decision Makers in Higher Education Institutions" (Unpublished doctoral dissertation). Robert Morris University. Retrieved November 19, 2018, from https://www.ecampus.edu.

Ali, O., J. Soar, M. J. Yong, and J. Biswas. 2015. "Anticipated Benefits of Cloud Computing Adoption in Australian Regional Municipal Government: An Exploratory Study." *Pacific Asia Conference on Information Systems*, 1–17.

Ali, O., J. Soar, M. J. Yong, and X. Tao. 2016. "Factors to Be Considered in Cloud Computing Adoption. *Web Intelligence* 14, no. 4: 309–323.

Alsanea, M., and D. Wainwright. 2014. "Identifying the Determinants of Cloud Computing Adoption in a Government Sector: A Case Study of Saudi Organization." *International Journal of Business and Management Studies* 6, no. 2: 29–43.

Angela, L. and C. Nan-Chou. 2012. "Cloud Computing as an Innovation: Perception, Attitude, and Adoption." *International Journal of Information Management* 32, no. 6: 533–540.

Asadi, S., M. Nilashi, C. H. Abd Razak, and E. Yadegaridehkordi. 2017. "Customer's Perspectives on Adoption of Cloud Computing in Banking Sector." *Information Technology and Management* 18, no. 4: 305–330.

Attaran, M., S. Attaran, and B. G. Celik. 2017. "Promises and Challenges of Cloud Computing in Higher Education: A Practical Guide for Implementation." *Journal of Higher Education Theory & Practice* 17, no. 6: 20–38.

Babbie, E. R. 2016. *The Practice of Social Research*, 14th ed. Boston, MA: Cengage Learning.

Baez, S. E., J. M. Hoch, and R. J. Cramer. 2019. "Social Cognitive Theory and the Fear-Avoidance Model: An Explanation of Poor Health Outcomes after ACL Reconstruction." *Athletic Training & Sports Health Care* 11, no. 4: 168–173.

Benlian, A., W. J. Kettinger, A. Sunyaev, T. J. Winkler, and Guest Editors. 2018. "Special Section: The Transformative Value of Cloud Computing; A Decoupling, Platformization, and

Recombination Theoretical Framework." *Journal of Management Information Systems* 35, no. 3: 719–739. https://doi.org/10.1080/07421222.2018.1481634.

Bernsteiner, R., and R. Pecina. 2015. "Cloud Based Office Suites: User Attitudes towards Productivity, Usability and Security" in International Conference on Knowledge Management in Organizations. 389–403.

Bhatia, S. 2014. "6 Surprising Ways Cloud Computing Is Changing Education." Retrieved from https://cloudtweaks.com/2014/12/cloud-computing-education-2/.

Brazier, A. 2014. "Organizational Change." *Loss Prevention Bulletin 239*, no. 1: 3–6.

Briscoe, Gerard, and Alexandros Marinos. 2009. "Digital Ecosystems in the Clouds: Towards Community Cloud Computing." In IEEE (corp. ed.). 2009 3rd IEEE International Conference on Digital Ecosystems and Technologies (DEST 2009): 103–108. New York, USA: Institute of Electrical and Electronics Engineers (IEEE).

Buabeng-Andoh, C. 2018. "Predicting Students' Intention to Adopt Mobile Learning: A Combination of Theory of Reasoned Action and Technology Acceptance Model." *Journal of Research in Innovative Teaching* 11, no. 2: 178–191.

Burfield, C. 2011. "Extending Face-to-Face Learning through Cloud Tools." *Distance Learning* 8, no. 4: 1–5.

Buyya, R., C. S. Yeo, and S. Venugopal, S. 2008. "Market-Oriented Cloud Computing: Vision, Hype and Reality for Delivering IT Services as Computing Utilities." Keynote paper, Proceedings of the 10th IEEE International Conference on High Performance Computing and Communications, Dalian, China, September 25–27.

Chandra, S., and K. N. Kumar. 2018. "Exploring Factors Influencing Organizational Adoption of Augmented Reality in E-Commerce: Empirical Analysis Using Technology-Organization-Environment Model." *Journal of Electronic Commerce Research* 19, no. 3: 237–265.

Chen, T., Ta-Tao Chuang, and Kazuo Nakatani. 2016. "The Perceived Business Benefit of Cloud Computing: An Exploratory Study." *Journal of International Technology & Information Management* 25, no. 4: 101–121.

Chipidza, W., and G. Green. 2019. "Why Do Students Not Major in MIS? An Application of the Theory of Planned Behavior." *Journal of Information Systems Education* 30, no. 2: 111–126.

Christensen, L. B., R. B. Johnson, and L. A. Turner. 2015. *Research Methods, Design, and Analysis*, 12th ed. Boston, MA: Pearson.

Chung, Y., and Y. Park. 2019. "What I Know, What I Think I Know, and Whom I Know." *Journal of Consumer Affairs* 53, no. 3: 1312–1349.

Chuttur, M. Y. 2009. "Overview of the Technology Acceptance Model: Origins, Developments and Future Directions, Indiana University, USA." *Sprouts: Working Papers on Information Systems* 9, no. 37. http://sprouts.aisnet.org/9-37.

CloudFoundryFoundation. 2012. "Cloud Foundry Application Runtime User Survey." https://cloud foundry.org/wp-content/uploads/2012/02/CFF_ApplicationRuntime_UserSurvey.pdf.

Cohen, J. 1988. *Statistical Power Analysis for the Behavioral Sciences*, 2nd ed. New York: Routledge.

Cohen, J., P. Cohen, S. G. West, and L. S. Aiken. 2002. *Applied Multiple Regression/Correlation Analysis for the Behavioral Sciences*, 3rd ed. New York: Routledge.

Compeau, D. R., and C. A. Higgins. 1995. "Computer Self-Efficacy: Development of a Measure and Initial Test." *MIS Quarterly* 19, no. 2: 189–211.

Cooper, D. R., and P. S. Schindler. 2014. *Business Research Methods*, 12th ed. Boston, MA: McGraw-Hill: Irwin.

Cresswell, J. W. 2009. *Research Design: Qualitative, Quantitative, and Mixed Methods Approaches*, 3rd ed. Los Angeles, CA: Sage Publications, Inc.

Daniel, C. 2018. "There's an (Updated) App for That." *Capitol Ideas* 61, no. 4: 10–13.

Davis, F. D. 1989. "Perceived Usefulness, Perceived Ease of Use, and User Acceptance of Information Technology." *MIS Quarterly* 13: 318–340.

Davis, F. D., R. P. Bagozzi, and P. R. Warshaw. 1989. "User Acceptance of Computer Technology: A Comparison of Two Theoretical Models." *Management Science* 35, no. 8: 982–1003.

———. 1992. "Extrinsic and Intrinsic Motivation to Use Computers in the Workplace." *Journal of Applied Social Psychology* 22, no. 14, 1111–1132.

Daylami, N. 2015. "The Origin and Construct of Cloud Computing." *International Journal of the Academic Business World* 9, no. 2: 39–45.

Deci, E. L., and R. M. Ryan. 1985. *Intrinsic Motivation and Self-Determination in Human Behavior.* New York: Plenum Press.

DeVellis, R. 2012. *Scale Development: Theory and Applications.* Los Angeles: Sage.

Dinjens, R. N., R. Senden, I. C. Heyligers, and B. Grimm. 2014. "Clinimetric Quality of the New 2011 Knee Society Score: High Validity, Low Completion Rate." *The Knee* 21, no. 3: 647–654.

Dillon, A., and M. G. Morris. 1996. "User Acceptance of New Information Technology: Theories and Models." *Annual Review of Information Science and Technology* 14, no. 4: 3–32.

Ďulík, M., & M. Ďulík Jr. 2016. "Security in Military Cloud Computing Applications." *Science & Military Journal* 11, no. 1: 26–33.

Effiong, A. 2020. "Small-to-Medium-Size Enterprise Managers' Experiences with Cloud Computing" (Order No. 27740266). Available from ProQuest Dissertations & Theses Global. (2392459741).

English, M. 2018. "Strategic and Managerial IT Issues: An Ongoing Business Challenge." *Strategic & Managerial IT Issues—Research Starters Business* 1, no. 1.

Etro, F. 2011. "The Economics of Cloud Computing." *The IUP Journal of Managerial Economics* 9, no. 2: 7–22.

Fishbein, M., and I. Ajzen. 1975. *Belief, Attitude, Intention, and Behavior: An Introduction to Theory and Research*. Reading, MA: Addison-Wesley.

Fitzgerald, S., J. Rumrill PD, and J. Schenker. 2004. "Correlational Designs in Rehabilitation Research." *Journal of Vocational Rehabilitation* 20, no. 2: 143–150.

Foster, Ian, Yong Zho, Ioan Raicu, and Shiyong Lu. 2008. "Cloud Computing and Grid Computing 360-Degree Compared." Grid Computing Environments Workshop (GCE), Austin, November 12–16: 1–10.

Gangwar, H., H. Date, and R. Ramaswamy. 2015. "Understanding Determinants of Cloud Computing Adoption Using an Integrated TAM-TOE Model." *Journal of Enterprise Information Management* 28, no. 1: 107–130.

Gartner. 2018. "Size of the Public Cloud Platform as a Service (PaaS) Market Worldwide from 2015 to 2020." https://www.statista.com/study/31311/platform-as-a-service-statista-dossier/.

George, D., and P. Mallery. 2010. *SPSS for Windows Step by Step: A Simple Guide and Reference 17.0 Update*, 10th edition. Boston: Pearson.

Georgescu, M., and M. Matei. 2013. "Small Steps for Cloud Computing towards the Future of Education." *ELearning & Software for Education* 2: 225–230.

Ghazizadeh, M., J. D. Lee, and L. N. Boyle. 2012. "Extending the Technology Acceptance Model to Assess Automation." *Cognition, Technology & Work* 14: 39–49. http://dx.doi.org/10.1007/s10111-011-0194-3.

Gibbons, R. 2019. "Counting the Costs of IaaS and SaaS." *Computer Weekly* 14. Retrieved from http://search.ebscohost.com/login.aspx?direct=true&AuthType=shib&db=f5h&AN=134950280&site=eds-live&scope=site.

Gill, R. 2011. "Why Cloud Computing Matters to Finance." *Strategic Finance* 92, no. 7, 43–47.

Gonçalves, V., and P. Ballon. 2011. "Adding Value to the Network: Mobile Operators' Experiments with Software-as-a-Service and

Platform-as-a-Service Models. *Telematics and Informatics* 28, no. 1: 12–21.

Gonevski, I. 2018. "Information Systems Managers' Attitudes toward Cloud Computing in US Companies" (Order No. 10828634). Available from Dissertations & Theses @ University of Phoenix (2064424552). Retrieved from https://search.proquest.com/docview/2064424552?accountid=35812.

Gonzalez, M. D., and M. L. & Smith Jr. 2014. "Are Cloud Computing Services Adoption Trends Changing?" *Franklin Business & Law Journal* 2014, no. 3: 120–144.

Gonzalez, N., C. Miers, F. Redígolo, T. Carvalho, M. Simplicio, M. Näslundy, M. and Pourzandiy. 2012. "A Quantitative Analysis of Current Security Concerns and Solutions for Cloud Computing." Escola Politécnica at the University of São Paulo (EPUSP), São Paulo, Brazil.

Gouws, T., and Rheede van Oudtshoorn. 2011. "Correlation between Brand Longevity and the Diffusion of Innovations Theory." *Journal of Public Affairs* 11, no. 4: 236–242.

Graham, C. M., and F. M. Nafukho. 2007. "Culture, Organizational Learning, and Selected Employee Background Variables in Small-Size Business Enterprises." *Journal of European Industrial Training* 31, no. 2: 127–144. https://doi.org/10.1108/03090590710734354.

Grossman, R. L., and Yunhong Gu. 2009. "On the Varieties of Clouds for Data Intensive Computing." *IEEE Computer Society Bulletin of the Technical Committee on Data Engineering* 32, no. 1: 44–51.

Grün, B., and S. Dolnicar. 2016. "Response Style Corrected Market Segmentation for Ordinal Data." *Marketing Letters* 27, no. 4, 729–741.

Gutierrez-Garcia, J. O., and K. M. Sim. 2013. "Agent-Based Cloud Service Composition." *Applied Intelligence* 38, no. 3: 436–464.

Habjan, K. B., and A. Pucihar. 2017. "Cloud Computing Adoption Business Model Factors: Does Enterprise Size Matter?" *Engineering Economics* 28, no. 3: 253–261. https://doi-org.contentproxy.phoenix.edu/10.5755/j01.ee.28.3.17422.

Hakim, Z. 2017. "Factors That Contribute to the Resistance to Cloud Computing Adoption by Tech Companies vs. Non-Tech Companies" (unpublished doctoral dissertation). Nova Southeastern University. Retrieved November 19, 2018, from https://www.ecampus.edu.

Hassan, H., M. Nasir, N. Khairudin, and I. Adon. 2017. "Factors Influencing Cloud Computing Adoption in Small and Medium Enterprises." *Journal of ICT* 16, no. 1: 21–41.

Heart, T. 2010. "Who Is Out There?: Exploring the Effects of Trust and Perceived Risk on SaaS Adoption Intentions." *ACM SIGMIS Database* 41, no. 3: 49–68.

Hsu, C. L., and J. Lin. 2016. "Factors Affecting the Adoption of Cloud Services in Enterprises." *Information Systems & E-Business Management* 14, no. 4: 791–822.

Huston, S. A. 2018. "Factors Associated with Entrepreneurial Intentions in Doctor of Pharmacy Students." *American Journal of Pharmaceutical Education* 82, no. 9: 1058–1072.

Ibrahim, H. M. 2014. "Assessing Cloud Computing Adoption by IT Professionals in Small Business Using the Technology Acceptance Model" (Order No. 3579718). Available from ProQuest Dissertations & Theses Global. (1510629794).

Iovan, S. and G. I. Daian. 2013. "Impact of Cloud Computing on Electronic Government." *Academica Brancusi* 1: 71–77.

Iqbal, S., M. L. M. Kiah, N. B. Anuar, B. Daghighi, A. W. A. Wahab, and S. Khan. 2016. "Service Delivery Models of Cloud Computing: Security Issues and Open Challenges." *Security & Communication Networks* 9, no. 17: 4726–4750.

Jawahar, D., and K. N. Harindran. 2016. "The Influence of Affect on Acceptance of Human Resource Information Systems with Special Reference to Public Sector Undertaking." *IUP Journal of Management Research* 15, no. 2: 33–52. Retrieved from https://search.proquest.com/docview/1789739766?accountid=45853.

Jiang, Y., D. Chen, and F. Lai. 2010. "Technological-Personal-Environmental (TPE) Framework: A Conceptual Model for Technology Acceptance at the Individual Level." *Journal of*

International Technology and Information Management 19, no. 3: 89–98.

Ji-Seong, J., M. Kim, Y. Kwan-Hee. 2013. "A Content Oriented Smart Education System Based on Cloud Computing." *International Journal of Multimedia & Ubiquitous Engineering* 8, no. 6: 313–327.

Jones, C. M. 2009. "Utilizing the Technology Acceptance Model to Assess Employee Adoption of Information Systems Security Measures" (doctoral dissertation). Retrieved from ProQuest Dissertations and Theses database.

Katzan, H. 2010. "The Education Value of Cloud Computing." *Contemporary Issues in Education Research* 3, no. 7: 37–42.

Kaur, A. 2018. "State of Mobile Cloud Computing within Government Sector." *International Journal of Advanced Research in Computer Science* 9, no. 3: 110–112.

Kerr, D. L. 1989. "Adoption of Competitive Advertising Practices among Professionals in a Deregulated Environment—A Test of Two Competing Models" (Order No. 8915752). Available from ProQuest Dissertations & Theses Global. (303763157). Retrieved from https://search.proquest.com/docview/303763157?accountid=45853.

Kim, H., 2017. "Enhancing Trusted Cloud Computing Platform for Infrastructure as a Service." *Advances in Electrical and Computer Engineering* 17: 9–14. doi: 10.4316/AECE.2017.01002.

Kim, S. 2012. "Factors Affecting the Use of Social Software: TAM Perspectives." *The Electronic Library* 30: 690–706.

Koğar, Esin Yilmaz, E. Demirdüzen, and S. Gelbal. 2017. "Cronbach's Coefficient Alpha: A Meta-Analysis Study." *Hacettepe University Journal of Education* 32, no. 1: 18–32.

Koul, S., and A. Eydgahi. 2017. "A Systematic Review of Technology Adoption Frameworks and Their Applications." *Journal of Technology Management & Innovation* 12, no. 4: 106–112.

Kraft, C. 2018. "What Is SaaS and How Can It Help Build Our Profession?" *Journal of Government Financial Management* 67, no. 2: 26–31.

Krancher, O., P. Luther, and M. Jost. 2018. "Key affordances of Platform-as-a-Service: Self-Organization and Continuous Feedback." *Journal of Management Information Systems* 35, no. 3: 776–812.

Krogh, S. 2018. "Anticipation of Organizational Change." *Journal of Organizational Change Management* 31, no. 6: 1271–1282.

Laudon, J. P., and K. C. Laudon. 2003. *Management Information Systems,* 7th ed. New Jersey: Prentice Hall.

Lee, Y., M. Li, T. Yen, and T. Huang. 2010. "Analysis of Adopting an Integrated Decision Making Trial and Evaluation Laboratory on a Technology Acceptance Model." *Expert Systems with Applications* 37, no. 2: 1745–1754.

Leedy, P. D., and J. E. Ormrod. 2016. *Practical Research: Planning and Design*, 11th ed. Boston, MA: Pearson.

Legris, P., J. Ingham, and P. Collerette. 2003. "Why Do People Use Information Technology? A Critical Review of the Technology Acceptance Model." Inf. Manag. 40, 191e204.

Levin, K. A. 2006. "Study Design III: Cross-Sectional Studies." *Evidence-Based Dentistry* 7: 24–25.

Lewis-Beck, M. S., A. Bryman, and T. Futing Liao. 2004. *The SAGE Encyclopedia of Social Science Research Methods.* Thousand Oaks, CA: Sage Publications, Inc. doi: 10.4135/9781412950589.

Li, J. 2016. "Technology Advancement and the Future of HRD Research." *Human Resource Development International* 19, no. 3: 189–191.

Lian, J. W., D. C. Yen, and Y. T. Wang. 2014. "An Exploratory Study to Understand The Critical Factors Affecting the Decision to Adopt Cloud Computing in Taiwan Hospital." *International Journal of Information Management* 34, no. 1: 18–36.

Lingens, B., S. Winterhalter, L. Krieg, and O. Gassmann. 2016. "Archetypes and Basic Strategies of Technology Decisions." *Research Technology Management* 59, no. 2: 36.

Low, C., Y. Chen, and M. Wu. 2011. "Understanding the Determinants of Cloud Computing Adoption." *Industrial Management & Data Systems* 111, no. 7: 1006–1023.

Madhavaiah, C., I. Bashir, and S. I. Shafi. 2012. "Defining Cloud Computing in Business Perspective: A Review of Research." *Vision (09722629)* 16, no. 3: 163–173.

Malley, A. 2015. "Licklider, Who Foretold a Digital Future, Showed What Science Can Do." Retrieved from http://link.galegroup.com/apps/doc/A418395539/GIC?u=uphoenix&sid=GIC&xid=0c3ab511.

Marković, D. S., I. Branović, and R. Popović. 2014. "Review of Cloud Computing in Business." *Singidunum Journal of Applied Sciences*: 673–677.

Marks, A. E., and B. Lozano. 2010. *Executive's Guide to Cloud Computing*. USA: John Wiley & Sons Inc., 25.

Marston S., Z. Li, S. Bandyopadhyay, J. Zhang, and A. Ghalsasi. 2011. "Cloud Computing—The Business Perspective." *Decision Support Systems* 51: 176–189.

McConnell, W. S. 2009. "Technology Adoption: Influence of Availability and Accessibility" (Order No. 3448397). Available from Dissertations & Theses @ University of Phoenix. (860120822). Retrieved from https://search-proquest-com.contentproxy.phoenix.edu/docview/860120822?accountid=35812

Meyvis, T., and S. M. J. Van Osselaer. 2018. "Increasing the Power of Your Study by Increasing the Effect Size." *Journal of Consumer Research* 44, no. 5: 1157–1173.

Misra, S. C., and A. Mondal. 2011. "Identification of a Company's Suitability for the Adoption of Cloud Computing and Modelling Its Corresponding Return on Investment." *Mathematical and Computer Modelling* 53, no. 3–4: 504–521.

Mollick, E. 2006. "Establishing Moore's Law." *IEEE Annals of the History of Computing* 28, no. 3: 62–75. doi:10.1109/MAHC.2006.45.

Muniasamy, V., I. M. Ejalani, and M. Anandhavalli. 2014. "Moving towards Virtual Learning Clouds from Traditional Learning: Higher Educational Systems in India." *International Journal of Emerging Technologies in Learning* 9, no. 9, 70–76.

Nikolova, M. 2012. "Cloud Computing in Government." Proceedings of the international conference on information technologies, 202–211.

Njenga, K., L. Garg, A. K. Bhardwaj, V. Prakash, and S. Bawa. 2019. "The Cloud Computing Adoption in Higher Learning Institutions in Kenya: Hindering Factors and Recommendations for the Way Forward." *Telematics and Informatics* 38: 225–246.

Nurmi, Daniel, R. Wolski, C. Grzegorczyk, G. Obertelli, S. Soman, L. Youseff, and D. Zagorodnov. (2008). "The Eucalyptus Open-Source Cloud-Computing System." *Cloud Computing and Its Applications*, CCGRID '09 Proceedings of the 2009 9th IEEE/ACM International Symposium on Cluster Computing and the Grid. Chicago: 124–131.

Obinkyereh, W. T. 2017. "Cloud Computing Adoption in Ghana: A Quantitative Study Based on Technology Acceptance Model (TAM)" (Order No. 10286333). Available from ProQuest Dissertations & Theses Global. (1914913372). Retrieved from https://search.proquest.com/docview/1914913372?accountid=35812.

Qiong, J. G. Yue, and S. J. Barnes. 2017. "Enterprise 2.0 Post-Adoption: Extending the Information System Continuance Model Based on the Technology-Organization-Environment Framework." *Computers in Human Behavior* 67, no. 1: 95–105.

Ojala, A., and P. Tyrvainen, P. 2011. "Value Networks in Cloud Computing." *The Journal of Business Strategy* 32, no. 6: 40–49.

Oliveira, T., and M. F. Martins. 2011. "Literature Review of Information Technology Adoption Models at Firm Level." *The Electronic Journal Information Systems Evaluation* 14, no. 1: 110–12.

Paquette, S., P. T. Jaeger, and S. Wilson. 2010. "Identifying the Security Risks Associated with Governmental Use of Cloud Computing." *Government Information Quarterly* 27, no. 3. 245–253.

Pardeshi, V. H. 2014. "Cloud Computing for Higher Education Institutes: Architecture, Strategy, and Recommendations for Effective Adaptation." *Procedia Econ. Finance.* 11: 589–599.

Park, J., and M. Park. 2016. "Qualitative versus Quantitative Research Methods: Discovery or Justification?" *Journal of Marketing Thought* 3, no. 1: 1–7.

Price, P., R. Jhangiani, and I. A. Chiang. 2015. *Research Methods in Psychology*, 3rd ed. Washington, DC: Saylor.org.

Rajaraman, V. 2014a. "Cloud Computing." *Resonance: Journal of Science Education* 19, no. 3: 242–258.

———. 2014b. "JohnMcCarthy—Father of Artificial Intelligence." *Resonance: Journal of Science Education* 19, no. 3, 198–207.

Rajendra Prasad, M., R. Lakshman Naik, and V. Bapuji. 2013. "Cloud Computing: Research Issues and Implications." *International Journal of Cloud Computing and Services Science* 2, no. 2: 134–140.

Ratten, V. 2013. "Exploring Behaviors and Perceptions Affecting the Adoption of Cloud Computing." *International Journal of Innovation in the Digital Economy* 4, no. 3: 51–68.

Raza, M. H., A. F. Adenola, A. Nafarieh, and W. Robertson. 2015. "The Slow Adoption of Cloud Computing and IT Workforce." *Procedia Computer Science* 52: 1114–1119.

Reese, G. 2009. *Cloud Application Architectures: Building Applications and Infrastructure in the Clouds*. Sebastopol, CA: O'Reilly Media.

Regalado, A. 2011. "Who Coined 'Cloud Computing'?" Retrieved from https://www.technologyreview.com/s/425970/who-coined-cloud-computing/.

Rittinghouse, J. W. and J. F. Ransome. 2016. *Cloud Computing: Implementation, Management, and Security*. Boca Raton, FL: CRC Press.

Rogers, E. M. 1995. "Diffusion of Innovations: Modifications of a Model for Telecommunications." *Die Diffusion von Innovationen in der Telekommunikation* 17: 25–38.

Ross, V. W. 2010. "Factors Influencing the Adoption of Cloud Computing by Decision Making Managers" (Order No. 3391308). Available from ProQuest Central. (305262031). Retrieved from https://search.proquest.com/docview/305262031?accountid=35812.

Roster, J., C. Moore, and K. Pfeiler. 2010. "Who Really Cares about the Cloud—An Industry Perspective." www.GartnerWebinars.gartner.com.

Rukh, L., M. A. Choudhary, and S. A. Abbasi. 2015. "Analysis of Factors Affecting Employee Satisfaction: A Case Study from Pakistan." *Work* 52, no. 1: 137–152.

Rzheuskiy, A., N. Veretennikova, N. Kunanets, and V. Kut. 2018. "The Information Support of Virtual Research Teams by Means of Cloud Managers." *International Journal of Intelligent Systems and Applications* 10, no. 2: 37.

Saini, L., and H. Kaur. 2017. "Role of Cloud Computing in Education System." *International Journal of Advanced Research in Computer Science* 8, no. 4: 345–347.

Sällberg, H., and L. Bengtsson. 2016. "Computer and Smartphone Continuance Intention: A Motivational Model." *The Journal of Computer Information Systems* 56, no. 4: 321–330.

Schenker, J. D., and P. D. Rumrill Jr. 2004. "Causal-Comparative Research Designs." *Journal of Vocational Rehabilitation* 21, no. 3: 117–121.

Schultz, B. 2011. "Public Cloud vs. Private Cloud: Why Not Both?" Retrieved from https://www.networkworld.com/article/2201101/public-cloud-vs--private-cloud--why-not-both-.html.

Seibold, David R., and Roy E. Roper. 1980. "Psychosocial Determinants of Health Care Intentions: Test of the Triandis and Fishbein Models." *Communication Yearbook* 3: 625–643.

Senarathna, I., C. Wilkin, M. Warren, W. Yeoh, and S. Salzman. 2018. "Factors that Influence Adoption of Cloud Computing: An Empirical Study of Australian SMEs." *Australasian Journal of Information Systems* 22. doi: https://doi.org/10.3127/ajis.v22i0.1603.

Sharifzadeh, M. S., C. A. Damalas, G. Abdollahzadeh, and H. Ahmadi-Gorgi. 2017. "Predicting Adoption of Biological Control among Iranian Rice Farmers: An Application of the Extended Technology Acceptance Model (TAM2)." *Crop Protection* 96, no. 1: 88–96.

Shawish A., and M. Salama. 2014. "Cloud Computing: Paradigms and Technologies." In *Inter-cooperative Collective Intelligence: Techniques and Applications*. Studies in Computational Intelligence, vol. 495, edited by Fatos Xhafa and Nik Bessis. Springer, Berlin: Heidelberg.

Shroff, R. H., F. S. T. Ting, and W. H. Lam. 2019. "Development and Validation of an Instrument to Measure Students' Perceptions of Technology-Enabled Active Learning." *Australasian Journal of Educational Technology* 35, no. 4. doi: 10.14742/ajet.4472.

Sivanadarajah, N., I. El-Daly, G. Mamarelis. M. Z. Sohail, and P. Bates. 2017. "Informed Consent and the Readability of the Written Consent Form." *Annals of the Royal College of Surgeons of England* 99, no. 8. 645-649. doi:10.1308/rcsann.2017.0188.

Sobragi, C. G., A. C. Gastaud Maçada, and M. Oliveira. 2014. "Cloud Computing Adoption: A Multiple Case Study." *Base* 11, no. 1: 75–91.

Solomon, F. 2009. "Input Identifies Future Cloud Computing for Government." *Policy & Practice* 3: 27.

Stanciu, D. I. 2017. "Are Current Models of Technology Acceptance Explanatory Enough? An Analysis of Potentially Underexploited Psychological Correlates of Technology Acceptance." *Elearning & Software for Education*: 2616–623. doi:10.12753/2066-026X-17-172.

Taherdoost, H. 2018. "A Review of Technology Acceptance and Adoption Models and Theories." *Procedia Manufacturing* 22: 960–967.

Thompson, R. L., C. A. Higgins, and J. M. Howell. 1991. "Personal Computing: Toward a Conceptual Model of Utilization." *MIS Quarterly* 15, no. 1: 124–143.

Tokunaga, H. T. 2016. *Introduction to Statistics: Fundamental Statistics for the Social and Behavioral Sciences*. Thousand Oaks, CA: SAGE Publications.

Tornatzky, L. G., M. Fleischer, and A. K. Chakrabarti. 1990. *The Processes of Technological Innovation*. Lexington, MA: Lexington Books.

Triandis, Harry C. 1977. *Interpersonal Behavior*. Monterey, CA: Brooks/Cole Publishing Company.

Tripathi, S., and N. Jigeesh. 2013. "A Review of Factors That Affect Cloud Computing Adoption." *The IUP Journal of Computer Science* 8, no. 4: 48–59.

Venkatesh, V., J. Y. L. Thong, and Xin Xu. 2016. "Unified Theory of Acceptance and Use of Technology: A Synthesis and the Road Ahead." *Journal of the Association for Information Systems* 17, no. 5: 328–376.

Wall Emerson, R. 2015. "Causation and Pearson's Correlation Coefficient." *Journal of Visual Impairment & Blindness* 36, no. 3: 242–244.

Wang, H., W. He, and F. Wang. 2012. "Enterprise Cloud Service Architectures." *Information Technology and Management* 13, no. 4: 445–454.

Weber, R. 2012. "Evaluating and Developing Theories in the Information Systems Discipline." *Journal of the Association for Information Systems* 13, no. 1: 2–30.

Wiley, D. C. and A. C. Cory. 2013. "Theory of Reasoned Action." In *Encyclopedia of School Health*. Thousand Oaks, CA: SAGE Publications, Inc. doi: 10.4135/9781452276250.n265: 646–647.

Yapp, E. H. T., C. Balakrishna, J. A. L. Yeap, and Y. Ganesan. 2018. "Male and Female Technology Users' Acceptance of On-Demand Services." *Global Business & Management Research* 10, no. 1: 105–126.

Zanello, G., X. Fu, P. Mohnen, and M. Ventresca. 2016. "The Creation and Diffusion of Innovation in Developing Countries: A Systematic Literature Review." *Journal of Economic Surveys* 30, no. 5: 884–912. https://doi.org/http://onlinelibrary.wiley.com/journal/10.1111/%28ISSN%291467-6419/issues.

ABOUT THE AUTHOR

With origins from the island of Jamaica, Dr. Lawrence was raised in Brooklyn, New York. After graduating high school, he joined the United States Air Force, where he served for over twenty-two years. During his time in the air force, he served in multiple combat zones and participated in just about every major operation since 1991, which included operations in Afghanistan, Africa, Bosnia, Croatia, Germany, Hungary, Iraq, Italy, Norway, Panama, and a host of other countries in the Southwest Asia region.

Since retiring from the military, Dr. Lawrence has worked as a defense contractor supporting the FBI, the Department of Defense, and a few corporate organizations. He has over thirty years of experience working in the information technology career field, having worked in the areas of IT service desk operations, IT service management, server operations, network management, enterprise compliance, architecture and engineering, and project management.

Dr. Lawrence is a graduate of the University of Maryland Global Campus with a bachelor's degree in computer science, a master's degree from Webster University in information technology management, and a doctorate in management with a concentration in information systems technology from the University of Phoenix. In addition to his educational accolades, he holds several industry certifications consisting of the PMP, ITIL v3, AWS Cloud Practitioner, Azure Fundamentals Practitioner, MS-ITP, C|HFI, Security +, Network +, and Project +.

He currently resides in Tampa, Florida, and operates his own IT consulting practice, LawTech IT Consulting, providing services in IT project management, ITSM, IT process improvement, and cloud integration solutions.

www.ingramcontent.com/pod-product-compliance
Lightning Source LLC
LaVergne TN
LVHW041035150826
845672LV00001B/337